AF817912

Spatial Interpolation for Climate Data

Spatial Interpolation for Climate Data

The Use of GIS in Climatology and Meteorology

Edited by
Hartwig Dobesch
Pierre Dumolard
Izabela Dyras

ISTE

First Published in Great Britain and the United States in 2007 by ISTE Ltd.

First South Asian Edition 2007

Apart from any fair dealing for the purposes of research or private study, or criticism or review, as permitted under the Copyright, Designs and Patents Act 1988, this publication may only be reproduced, stored or transmitted, in any form or by any means, with the prior permission in writing of the publishers, or in the case of reprographic reproduction in accordance with the terms and licences issued by the CLA. Enquiries concerning reproduction outside these terms should be sent to the publishers at the undermentioned address:

ISTE Ltd.
6 Fitzory Square
London WJT 5DX
UK

ISTE USA
4308 Patrice Road
Newport Beach, CA 92663
USA

www.iste.co.uk

© ISTE Ltd., 2007

The rights of Hartwig Dobesch, Pierre Dumolard and Izabela Dyras to be identified as the authors of this work has been asserted by him in accordance with the Copyright, Designs and Patents Act 1988.

This edition is for sale in the Indian subcontinent only. Not for export elsewhere.

ISBN 10: 1 905209 70 3
ISBN 13: 978 1 905209 70 5

Library of Congress Cataloging-in-Publication Data

Spatial interpolation for climate data: the use of GIS in climatology and meteorology/
edited by Hartwig Dobesch, Pierre Dumolard, Izabela Dyras.
p. cm.
Includes bibliographical references and index.
ISBN 10: 1 905209 70 3
ISBN 13: 978 1 905209 70 5
1. Climatology--Data processing. 2. Meteorology--data processing 3. Geospatial data--mathematical models. 4. Geographic information systems. 5. Spatial data infrastructures. I. Dobesch, Hartwig. II. Dumolard, Pierre. III. Dyras, Izabela.
QC874.3.S63 2007
551.60285--dc22

2007012743

British Library Cataloguing-in-Publication Data
A CIP record of this book is available from the British Library
ISBN 978 1 905209 70 5

Printed and bound in India by Brijbasi Art Press Limited, I-72, Sector-9, Noida, U.P., India.

Table of Contents

Preface

The COST 719 European research program (*"The use of GIS in climatology and meteorology"*) began in 2001 and ended in 2006. 20 European countries participated. Its main objective was to establish interfaces between GIS and data in climatology and meteorology and in order to reach this objective, three working groups were defined:

– working group 1 on "Data access and data availability";

– working group 2 on "Spatial interpolation";

– working group 3 on "GIS applications".

Most of the applications have focused on three climate parameters: temperature, precipitation and energy balance in the fields of climatology, meteorology and environmental sciences.

This book is the proceedings of most of the presentations made during the final conference of the COST 719 action (Grenoble, July 2006). It comprises four parts, each one introduced by a keynote speaker.

Part 1 is devoted to GIS use in meteorology and climatology. It is introduced by a text underlining that GIS is a mature technology to integrate, analyze and display spatial data but is still scarcely used in most Meteorological Services, partly due to the lack of an atmospheric data model (meteorology has its own operational infrastructure). A major concern is the share of terrain data in Europe (INSPIRE initiative) and the definition of a common metadata standard.

Four other contributions complete this first part.

– SIGMA is a currently running GIS (in Brazil) dedicated to real-time information (precipitation, temperature, lightning, NDVI, ozone, etc.) and alerts.

– A paper on webmapping of climatological and meteorological data compares ARC-IMS, open standards and Scalable Vector Graphics approaches.

– MISH (*Meteorological Interpolation method*) has been developed in Hungary. Geostatistical methods are based on one realization whereas MISH incorporates, in order to model the statistical parameters, climatic spatio-temporal information.

– A simple GIS study of the northern French Alps meteorological network shows that its statistical robustness is not evenly distributed, so that the sample should be stratified and the uncertainty regionalized.

Part 2 is dedicated to the spatial interpolation of climatological parameters. It is introduced by a chapter showing the developments in spatialization of meteorological and climatological parameters. Interpolation methods, which are present in most GIS, make it possible to combine numerous layers to derive estimates of parameters for any place at any time.

– The first application, in Poland, tests several methods for interpolating air temperatures, precipitation and cloudiness. The results for the temperature are correct, whereas the two other ones are in fact much more complicated to interpolate.

– Two following applications deal with the spatialization of (mean or average) temperatures in the northern parts of the French Alps. The measuring network is dense but still not dense enough to derive very good estimates for any place at any time (especially with anticyclonic weather).

– The last application discusses several methods for the spatialization of the radiation balance. It also examines the possibility to derive land surface albedo from satellite images.

Part 3 is devoted to demonstration projects. It is introduced by an overview of ready-to-use demos. The chapter insists on the need for relevant meteorological and climatological data for environmental users, presenting links and connections between simulation models developed by ecologists and hydrologists on the one hand, applications by meteorologists and climatologists on the other hand.

– The first demo deals with daily fire risk mapping in Portugal, combining structural elements (number of fires, burnt areas, vegetation, biomass accumulation, climatic variables) with weather forecasts for the next 24 and 48 hours.

– The second demo compares, in Spain, a high resolution precipitation database with one created by dynamic downscaling; the statistical analysis shows a good similarity in terms of spatiotemporal distribution and total precipitation.

– The third demo is devoted to simulating drought sensitivity in southern Hungary under the hypothesis of climate change and assessing that damages will depend more on vulnerability than on events themselves.

– The fourth demo presents the ALP-IMP project where the monthly temperature fields of the Greater Alpine Region were calculated back to 1760. In the new ECSN-HRT-GAR project, monthly climatology fields are modeled according to quality checked normals for about 1,700 stations. In its final stage, monthly grids (1 km^2 pixels) will be produced.

– The fifth demo presents the RWIS (Road Weather Information System) for the winter maintenance of roads. Road weather forecasts use sensors, sky view factor analysis and mesoscale weather forecasts. An energy balance model (IceMiser) has been tested on 12,000 locations of 6 roads in England.

Part 4 is dedicated to environmental problems, which are strongly related to climate. It is introduced by a chapter presenting the challenges in fine scale applications, with examples in agrometeorology and urban climatology.

– In regions (like Brittany) with an intensive agriculture, transfer of pollutants into water resources partly depends on land use management during winter (bare or cultivated soils). A predictive model at the piece of land scale, based on the Dezert-Smarandache theory, proved an 82% success rate.

– The DEMETER project has been set up to facilitate the water management (quantity and quality) for irrigated farming in southern Europe. Pilot studies have been carried out using agricultural surveys, satellite images and weather and climate data, showing the possible improved management of cultures in Mediterranean areas.

– Olfactive nuisances around landfill sites are usually associated with certain meteorological situations. To evaluate the population exposure, a metamodel has been built, combining a local meteorological model (ARPS with nested domains), terrain data and a specific Eulerian dispersion model.

– The last application deals with the estimated disaggregation of N_2O fluxes due to agricultural soils in Italy (N_2O has a warming potential 275 times greater than CO_2 and has a very long life). The disaggregation procedure was conducted with land use, environment and climate data.

Hartwig Dobesch
Pierre Dumolard
Izabela Dyras

Part 1

GIS to Manage and Distribute Climate Data

Chapter 1

GIS, Climatology and Meteorology

1.1. GIS technology and spatial data (working group 1)

1.1.1. *Introduction*

In the framework of the COST program, COST-719 addresses the spread of knowledge and skills concerning Geographical Information Systems (GIS) or, specifically, spatial data management within the climatological and meteorological community. GIS can offer a practical and relevant working environment for the integration, analysis and visualization of this data together with other spatial data sources. Within most National Meteorological Services (NMS) the acceptance of commercial GIS tools beyond climatology is still a cumbersome process, which is partly caused by the shortcomings underlying the data model (time aspects!) and partly by the lack of knowledge of applicable GIS methods. Another reason is that atmospheric science is "more concerned with the question of why phenomena happen and less with the region where they happen" [PET 01].

Geographical data can further enhance atmospheric applications. Nowadays, it appears to be quite difficult to collect uniform geographical data such as Digital Terrain Models and land cover on a European scale with a sufficient quality and affordability. Therefore, recent developments, such as the GMES initiative of the EC and ESA, the publication of the European Directive on public access to environmental information (28 January 2003) and the Water Framework Directive, are of the utmost importance for the user community that would benefit from the resulting information and accessible data sources.

Chapter written by Antonio PERDIGAO.

COST-719 aims to gather and disseminate know-how and technical skills among the different member states in order to increase the use of GIS tools for the management and integration of climatological, meteorological and other environmental data obtained from different sources.

The action is structured into working groups, one of which deals with the so-called watchdog function, searching for publicly accessible European spatial databases and the inventory and dissemination of knowledge about relevant software developments.

1.1.2. *Weather and GIS*

NMSs do not require such applications. Another could be that the applications are best suited to the needs of the NMS users. From the number of "GIS-like" tools present in most weather services it appears that the first conclusion is false. The data used to study weather and climate is spatial and therefore typically processed in some kind of spatial information system. The major problem exists in the time dimension of the meteorological data. Commercial GIS packages are only starting to deal with spatio-temporal data models which are relevant to atmospheric data. Christakos *et al.* [CHR 02] state in their introduction that: "...commonly used temporal geographical information systems neglect essential dynamics of the natural processes in time and do not take into account important cross-correlations and causal dependencies in the composite space/time domain...". As this development will take some time to mature, a more pragmatic approach should help to interface GIS and Atmospheric Science Information Systems (ASIS). Nativi *et al.* [NAT 04] describe the differences between both underlying data models and advocate models that are supported by so-called interoperability services. This is extremely important as traditional GIS users are, for example, willing to incorporate weather information in their applications (radar data for hydrologists) whereas meteorological researchers are interested in local topography for their high-resolution models (downscaling).

Shipley *et al.* [SHI 96] already noted that "...GIS does the weather", meaning that COTS-GIS packages are able to deal with the basic functionality that is also available in expensive, dedicated weather processing systems and with more open and standardized packages, this is even truer. These developments notwithstanding, GIS plays a modest role in the atmospheric research stage during which researchers tend to develop their own tools.

1.1.3. *Geographical data, environmental data and weather data*

In meteorology and climatology, geographical data plays a key role. The demand for these kinds of data can be considered from two aspects:

– They are used in analyses, for various kinds of spatial interpolation and modeling, in which case the data are a factor that determine the spatial distribution of the analyzed meteorological phenomena or climatological elements. This is the case for derivation of parameters for traffic safety applications (e.g. fog, glazed frost, heavy rainfall, storm), NWP modeling (description of physiographic features that influence surface fluxes), impact studies based on air pollution dispersion models, etc.

– They are used for visualizing the results of measurements or calculations, in which case they are used mainly as a background image for other data. Meteorological modeling takes advantage of climatic data, whereas climatic analyses are based on meteorological data and in both fields geographical data constitutes vital input information. In this case, geographical data concerns mainly elevation (with slope and exposure of the area), land cover (with parameters dependent on and independent of the season of the year), hydrography and soils.

This requires not only synoptic data about weather elements from many years, but also data about the geographical environment. This data significantly influences spatial distribution of meteorological elements that are shaped and influenced by geographical factors. Mainly, but not solely, the applied data pertains to the topographic profile. For instance, air temperature is clearly correlated with the height above sea level, distance from water reservoirs, slope, longitude and latitude.

Geographical data changes very slowly over time with respect to meteorological data. Therefore, in operational models, e.g. weather forecasting, they are treated differently from current measurements. Once prepared and processed, geographical data can still be used even for a few years. It is important though, to allow for the changeability of certain geographical elements in respective seasons of the year, or even months (e.g. leaf area index, seasonal changes in vegetation).

An aspect of the application of geographical data in meteorology and climatology is the presentation of measurements or the results of models and analyses to a range of end users. This is well exemplified by weather forecast maps and satellite or radar images, which are presented, for example, on Web pages of meteorological services. In this case, geographical data is mainly used as land-sea masks, reference maps or simply as background images appealing to our sense of aesthetics. Data about administrative divisions, location of places, elevation or hydrography are very helpful for both advanced users of operational meteorological products and the general public in order to understand the nature of a phenomenon.

One of the prerequisites to successfully embed GIS in a NMS is the availability of geographical data crossing national borders. While the acccss to weather data has always been clearly defined for third parties, i.e. non-NMSs using data in XML or GIS formats, finding accurate, cost-efficient and up-to-date geographical data is much harder. Pan-European datasets are particularly difficult to obtain, especially if a particular fitness-for-use is required.

Recently, such data has started to become available, for example:

– Height data. Some countries have a mapped topography at a high resolution resulting in very precise height maps. A European alternative is the dataset gathered in the framework of the Shuttle Radar Topography Mission (SRTM), with data points approximately every 30 meters.

– Land cover data. At national scales good datasets are available at a price. The CLC2000 project pursues a European Land Cover map based on a satellite image classification and will be available for free to registered users.

The INSPIRE initiative relating to the European spatial data infrastructure is very important for the future of data management. By considering the classification used under the INSPIRE initiative, the geographical data used in the meteorological and climatological realm is generally basic (reference) data and its availability at the European level should increase. Conversely, atmospheric conditions and meteorological spatial features are classified as thematic data and are also the focus of interest of INSPIRE and related initiatives (like GMES) which will help resolve interoperability problems.

1.1.4. *A GIS approach to access weather data*

Meteorology has its own data infrastructure based on the Global Telecommunication System (GTS) which is operating according to a so-called push mechanism. Station observations, for example, are shared worldwide within the hour in order to obtain a synoptic view of the current weather conditions. Therefore, projects such as UNIDART have been established within the framework of EUMETNET, which is the network of European NMSs.

This industry offers increasingly promising solutions, as is the case in the interesting development of ESRI on the establishment of ArcWeb products, which enable the data to be used in ArcGIS or some custom-made application. A good example comes from Meteorlogix, a US enterprise that offers various meteorological services to the public, containing information about cloud cover, NEXRAD (weather radar) base reflectivity and precipitation type (rain, mix, snow) and surface observations.

Now that this product complies with the standards of the Open GIS Consortium (OGC) regarding map servers, communication with other non-commercial, although OGC-compliant, software is guaranteed.

If a NMS has to open up part of its data collection, for example as a result of a national or European policy, it should preferably link up to a European infrastructure initiative. However, due to the status of construction of the projects, a more pragmatic approach is often followed, meaning that a lot of effort is placed on temporary solutions (dedicated databases or map servers). It is advocated that once such an approach is followed, the compliance to standards and extensible sources is kept in mind.

1.2. Data and metadata

1.2.1. *Introduction*

The need for the correct usage and improvement of a data infrastructure for climatology and meteorology is related to the availability and quality of data and relevant policies, taking into account that some European national meteorological services (NMSs) operate in a semi-commercial way.

The European strategy for supplying cartographical or thematic maps with topographical references is not the same for all countries where the scale, accuracy and precision can differ considerably. Data availability is sometimes restricted or very expensive resulting in the usage of free data which does not bear the necessary accuracy and precision, as is the case for GTOPO30 data.

1.2.2. *Important datasets*

The following provides an overview of datasets and/or projects that are relevant for climatology and meteorology.

Land use/land cover

The *CORINE Land Cover* project aims to provide a satellite image snapshot of Europe (*IMAGE 2000*), an up-to-date land cover map for the year 2000 (*CLC2000*) and information on mainland cover changes in Europe during the period 1990-2000. The satellite data is mainly derived from Landsat 7 Enhanced Thematic Mapper. CORINE is a joint initiative of the European Environment Agency (EEA) and the Joint Research Centre (JRC).

In addition to the standard products a number of derived outputs are produced (mainly aggregated or generalized datasets such as land cover grid maps at 100 m, 250 m and 1 km grid cell size). The data can be downloaded with no charge after a registration process.

Pelcom monitors the European Land Cover and is based on the interpretation of multispectral and multitemporal 1-km resolution NOAA-AVHRR satellite data. Contrary to I&CLC2000, Pelcom is not currently updated (last update 2000). The data can be acquired free of charge from the Website in ESRI Grid, Imagine or ENVI format.

Height

The *GTOPO30* height datasets distributed by the USGS are common due to the fact that usage is free and the data is easy to process within a GIS environment. However, the datasets are far from consistent and their application should be considered with care.

The *ACE* database is a global digital elevation model at 30 arc-seconds resolution generated to a large extent by means of satellite altimetry. De Montfort University in Leicester (UK) has developed it.

The *Global Land One-Km Base Elevation* (*GLOBE*) project provides a digital dataset for free which is considered more accurate than GTOPO30 and has an ongoing process of data collection. GTOPO30 and GLOBE are comparable but produced independently.

This data was collected aboard the NASA Space Shuttle Endeavour Radar Topography Mission (*SRTM*) (February 11-22, 2000). European data is available at a resolution of 3 arc-seconds (approximately 100 meters at the equator).

Environmental information systems

The *UNEP GRID* database is maintained for the purpose of assisting the international community and individual nations in making sound decisions with respect to resource management and environmental planning. Within the overall GRID-network, GRID-Geneva focuses on the acquisition, creation, documentation, archive and dissemination of global and European digital georeferenced environmental data.

Terrestrial Environment Information System (*TERRIS*) is a GIS that supports spatial data related activities of the European Topic Centre on Terrestrial Environment (ETCTE) of the EEA. The ETCTE Spatial Analysis Group's main tasks include the following:

– environmental spatial data creation and correction (such as CORINE);

– spatial data management, data collection, harmonization and distribution;

– map production for specific products;

– spatial analysis for specific projects;

– development of an environmental indicator.

TERRIS is expected to have the following characteristics:

– includes the terrestrial georeferenced information held by the EEA, in a predetermined format, under a single reference system: datum ETRS89, ellipsoid GRS80;

– all the geographic databases are documented by using a simplified metadata version of ISO TC211 19115;

– all information (databases linked to geographic coverage) is stored appropriately (using a DBMS as required), such as to facilitate access and use.

1.2.3. *Metadata*

Within the World Meteorological Organization (WMO) of the United Nations, the role of ISO 19115 is stressed as the metadata standard for describing meteorological and climatological data. Although this standard has been developed for the geographical community, it is useful for meteorology as they both deal with spatial data.

Within the framework of metadata it is necessary and relevant to know the following:

– The priority information (fields) within a metadata record, which present a profile for thematic data to be used.

– The level of metadata creation, how metadata is to be created for downloading datasets, or establishing coordinate reference systems and file formats.

– The tools for establishing metadata, including tools linked to GIS software and other freeware tools.

– The formats and systems for the dissemination of metadata.

Metadata provides information about data. Both intrinsic and extrinsic metadata is relevant as the former refers to the qualities of the dataset, e.g. accuracy and lineage, whereas the latter refers to the context of the dataset, e.g. access rules, costs, etc. In this data infrastructure metadata databases are required to find and evaluate data on a network such as the Internet.

For dynamic data such as atmospheric data a sophisticated classification system is required. Spatial data infrastructures can benefit from standardized DB queries as the metadata catalogs are similarly structured.

The INSPIRE Directive defines "metadata" as information describing spatial data services that make it possible to discover, list and use them. The information needed to create metadata is often readily available when the data are collected. The initial expense of documenting data clearly outweighs the potential costs of duplicated or redundant data generation.

1.2.4. *Open Geospatial Consortium*

The Open Geospatial Consortium (OGC) is a not-for-profit collaboration of leading industries running an international consensus process related to metadata, storage formats and interoperability of data. It includes most of the world's geoprocessing software vendors as well as integrators, universities and government agencies, which agree on open, standard software interfaces that enable interoperation between geoprocessing systems from different vendors and various kinds (e.g. GIS, remote sensing, automated mapping and facilities management, etc.).

1.2.5. *EU strategies for data handling and standards*

INSPIRE

The objective of INSPIRE is to improve the quantity and quality of data available for the Community policy-making and implementation in the Member States at all levels, and increase the availability of relevant, harmonized and high quality geographic data at EU level to support the formulation, implementation, monitoring and evaluation of community policies.

Detailed spatial information is available in Europe to support a broad range of policies. Indeed map-based information is used in many reporting, analyzing, evaluating and forecasting tools and activities. The main problems relate to data gaps, missing documentation, incompatible spatial datasets and services due to, for example, varying standards and barriers to the sharing and reuse of spatial data.

Pressures and impacts on the environment often cross national borders. Environmental policies therefore require the establishment of environmental management, as is the case for river basin districts established under the Water Framework Directive which require interoperable spatial information across national borders. The proposed INSPIRE Directive will provide consistent Community-wide presentation of spatial data and data quality, more Community-wide spatial information of greater consistency, integrated Community-wide services and rules to find and access this information.

The evolving INSPIRE GeoPortal intends to be Europe's Internet access point for spatial data and services under the Infrastructure for Spatial InfoRmation in Europe (INSPIRE) initiative, which aims to provide the functionalities needed for spatio-temporal data discovery, access and services, including searching, viewing and publishing metadata and data from numerous data suppliers through the Internet.

GMES

The initiative of Global Monitoring for Environment Security (GMES) is an EC-ESA joint strategy to establish a European capacity to provide operational information for monitoring and management of the environment and for civil security by 2008.

EIONET has been identified by the GMES initial period report as a priority network for the implementation of GMES as well as the European shared information system.

Land monitoring is one of three GMES pilot services identified by the European Commission and the European Space Agency (ESA) as a priority service to be developed within GMES.

The land monitoring services proposed will support a wide range of environmental information needs in Europe and build on the results and experiences from initial GMES projects, existing elements of operational land monitoring activities in Europe and the prioritizing criteria of the GMES Action Plan 2004-2008.

The *Water Framework Directive* (*WFD*), which was published and came into force on 22 December 2000, is a legislative framework to protect and improve the quality of all water resources such as rivers, lakes, groundwater, transitional and coastal water resources within the European Union. In order to achieve a "good status" for all European waters by 2015, a set of activities must be carried out, such as identifying individual river basins and assigning them to river basin districts, determining and characterizing groundwater and surface water bodies by establishing protected areas, setting up water status monitoring networks, renewing the impact of human activity.

As part of the Common Implementation Strategy for the WFD the key points proposed by the GIS-WG should lead to the creation of SDIs that match the GIS requirements of the WFD, in line with the INSPIRE initiative principles.

GINIE (Geographic Information Network in Europe) was a research project funded by the Information Society Technology Program of the EU and ended in January 2004. Partners were EUROGI, which is the European Umbrella Organization for Geographic Information, the Open Geospatial Consortium Europe representing the Geographic Information (GI) industry, the JRC of the European Commission and the University of Sheffield (coordinator).

The *EUFOREO* (EU Forum on Earth Observation Use for Environment and Security) project is a Thematic Network (TN) funded by the EC under the RTD Fifth Framework Program. The purpose of the EUFOREO network is to develop an efficient implementation of EU and national policies for Environment and Security through the use of Earth Observation (EO) technology.

EUFOREO participants include 18 members, among others ESA, JRC, EUMETSAT, EARSC and EUROSPACE.

DEMIS, the Open Geospatial Web Map Server (WMS) protocol, defines a simple interface for Web-based mapping applications. The WMS protocol is based on a simple query syntax for posting a request for the desired layers and zoom window to the server, which returns a map as a standard picture.

1.2.6. *Meteorological datasets, important projects and programs*

The *EUMETSAT* Geostationary satellite program includes the continuation of the current Meteosat system with the launch of the Meteosat Second Generation (MSG). Currently, two of EUMETSAT's new generation satellites have been launched into space.

On 7 October 2006 a EUMETSAT polar satellite (EPS/Metop-1), which is the first European polar orbiting operational meteorological satellite, will be launched. This is the first satellite series that will greatly enhance the range of observations made by EUMETSAT, particularly in support of climate and environment monitoring.

The development of new products and services through the concept of the Satellite Application Facilities (*SAF*) is evolving. Currently there are eight SAFs: 7 in pre-operational phase and 1 in development phase.

The *EUMETNET-UNIDART* (Uniform Data Request Interface) program has as its main goal the development of a Web-based information system that allows for a uniform and integrated access to heterogenous and distributed data sources which store all types of meteorological data and products. This goal is achieved by the integration of different data sources whose data and products are stored in different formats and locations.

UNIDART may be used as a standardized Web interface to the national climate database by using advanced technology (DATAGRID) for a secure access which may be used for other projects. It is in accordance with the development of WMO Information Systems (WIS), assumed as a single coordinated global infrastructure, intended to serve all the relevant WMO programs and aims to increase the efficiency of the meteorological hydrological societies.

Unidata has a mission to help users of earth-related data. Most of the data is provided in "real-time" or "near-real-time" revealing information through an Internet Data Distribution (IDD) system.

The Unidata community is building a system for disseminating near-real-time earth observations via the Internet. Unlike other systems, which are based on data centers where the information can be accessed, the Unidata IDD is designed so that users can request certain datasets.

A global community of netCDF users includes data providers, software developers and researchers who need portable, self-describing data. Use of netCDF for model outputs and data archives can enhance interoperability and make distributed data access practical.

1.2.7. *Projects using Earth Observation satellites*

There are currently 68 Earth Observation satellite missions operating and around 100 more missions, which carry over 300 instruments, planned for operation during the next 15 years around the world.

The meteorological Global Observing System encompassing both space-based and *in-situ* measurements contributes to two GMES objectives: global climate monitoring and the forecast of meteorological risks. Within the EUCOS (EUMETNET Composite Observing System) Program, which is supported by 18 European Meteorological services, it was recently decided to address the scientific aspects of optimal observing system design and downstream forecast of severe weather, as well as the development of end-to-end services on a European scale: Global Earth Observation System of Systems (GEOSS).

GEOSS will build on existing Earth Observation systems supporting interoperability, sharing information and reaching a common understanding of user requirements, supporting different areas as disaster mitigation and management, health, energy, climate, water, weather, ecosystems, agriculture and biodiversity. Under the GEOSS plan, satellites and *in-situ* observations from different countries/regions will be linked and participants will share access to Earth Observation data and model inputs.

The objective of future missions planned by the CEOS (Committee on Earth Observation Satellites) agencies over the next decade is the provision of improved operational observations for meteorology or new research capabilities, including:

– finer spatial temporal, and spectral measurements of atmospheric parameters, which enable a more accurate determination of parameters such as temperature and moisture. This data which is acquired from missions such as Aqua (NASA), METOP (EUMETSAT) and NPOESS (NOAA) is expected to lead to substantial improvements in the accuracy of mid- and long-range weather forecasts;

– the introduction of a new series of polar orbiting missions to complement the current NOAA series: the METOP series (EUMETSAT) was launched in October 2006 and the FY-3 series (China) soon afterwards;

– new capabilities for monitoring precipitation and cloud properties: by 2007, a constellation of satellites will be in place (comprising Aqua, Aura, CALIPSO, Cloudsat, PARASOL and OCO) and will fly in orbital formation (known as A-Train) to gather data which is needed to evaluate and improve the way clouds are represented in global models. The Global Precipitation Mission (GPM) will provide global observations of precipitation every three hours;

– soil moisture measurements which are important for the initialization of NWP models will be provided by SMOS (ESA);

– global three-dimensional wind fields: direct measurements will be made from space for the first time in 2007 by the ADM Aeolus mission (ESA) with the aim of improving weather forecasting and climate research;

– the IGOS (Integrated Global Observing Strategy) partnership provides a forum for establishing the performance and timing necessary from CEOS agency missions in order to satisfy the information requirements of the IGOS themes and of international programs such as the Global Climate Observing System (GCOS), Global Ocean Observing System (GOOS), the Global Terrestrial Observing System (GTOS), the World Climate Research Program (WCRP) and the International Geosphere-Biosphere Program (IGBP).

1.3. Interoperability

1.3.1. *Introduction*

For many years now, at various institutions and organizations, data has been collected and tools for its processing have been developed. Frequently, the same data and tools with identical functionality are collected and developed in different places, which increases costs and can be a cause of incompatibility in information obtained from various sources. This problem has been recognized for a long time, and the proposed solution with respect to spatial information is the Spatial Data Infrastructure (SDI).

The GSDI defines SDI as follows.

Spatial data infrastructures (SDIs) provide a basis for spatial data discovery, evaluation, and application, and include the following elements:

– *Geographic data*: the actual digital geographic data and information.

– *Metadata*: the data describing the data (content, quality, condition and other characteristics). It enables structured searches and comparison of data in different clearinghouses and gives the user adequate information to find data and use it in an appropriate context.

– *Framework*: it includes base layers, which will probably differ from location to location. It also includes mechanisms for identifying, describing and sharing the data by using features, attributes and attribute values, as well as mechanisms for updating the data without complete re-collection.

– *Services*: to help discover and interact with data.

– *Clearinghouse*: to actually obtain the data. Clearinghouses support a uniform, distributed search through a single user interface; they allow the user to obtain data directly, or they direct the user to another source.

– *Standards*: created and accepted at local, national and global levels.

– *Partnerships*: the glue that holds it together. Partnerships reduce duplication and costs of collection, as well as leveraging local/national/global technology and skills.

In Europe, the idea of SDI is implemented in practice as part of the INSPIRE initiative described in more detail in section 1.2.5, which also contains information about spatial databases available at the Europe-wide level. SDI is strongly linked to the concept of interoperability and it is the technical aspects of this concept which is the subject of the present chapter.

According to the definition provided by the US Institute of Electrical and Electronics Engineers, adopted by INSPIRE, *interoperability* is the ability of two or more systems or components to exchange information and to use the information that has been exchanged. In other words, interoperability is the ability to combine information and functionality from different systems implementing data sharing.

Interoperability is associated with the idea of modular solutions which enable the cooperation of components which are independent of the programming languages used to build them, as well as of the hardware platforms and operating systems on which they have been implemented.

From a technical point of view, interoperability is based on standards and technological solutions. The role of ISO TC 2111 (ISO 19100 family of standards) and the Open Geospatial Consortium (OGC specifications) is crucial with respect to the standardization in the field of digital geographic information. Among the technological solutions, Service-Oriented Architecture (SOA) is one of the most promising. The proposed technology for spatial data discovery, visualization and access is based on geospatial Web services, metadata Standards and XML implementations.

1.3.2. *Technology for service-oriented architectures*

The term SOA expresses a perspective of software architecture that defines the use of services to support the requirements of software users. In an SOA environment, the nodes on a network make resources available to other participants in the network as independent services that the participants access in a standardized way (i.e. by using Simple Object Access Protocol). The SOA comprise loosely coupled Web services which interoperate according to a formal definition which is independent from the underlying platform and programming language (e.g. Web Services Description Language (WSDL)). The interface definition encapsulates the vendor and language-specific implementation. An SOA is independent of development technology (such as COM, NET or Java).

The SOA is closely related to the open protocols, such as Hypertext Transfer Protocol (HTTP), Extensible Markup Language (XML) and Simple Object Access Protocol (SOAP), WSDL and Universal Description Discovery & Integration (UDDI).

WSDL is often used in combination with SOAP and XML Schema to provide Web services over the Internet. A client program connecting to a Web service can read the WSDL to determine what functions are available on the server. Any special data types used are embedded in the WSDL file in the form of XML Schema. The client can then use SOAP to actually call up one of the functions listed in the WSDL.

SOAP is a communication protocol for communication between applications which defines a format for sending messages between communicating applications via the Internet and specifically by using HTTP. SOAP is platform-independent and language-independent, and SOAP messages are encoded using XML. This means that SOAP provides a way to communicate between applications running on different operating systems, with different technologies and programming languages.

Based on a common set of industry standards, including HTTP, XML, XML Schema and SOAP, UDDI provides an interoperable, foundational infrastructure for a Web services-based software environment for both publicly available services and services which are only exposed internally within an organization. UDDI builds on the vision of a "meta-service" for locating Web services by enabling robust queries against rich metadata. It offers the industry a specification for building flexible, interoperable XML Web services registries which are useful in private as well as public uses.

1.3.3. *Interoperability in GIS*

"Firstly, 'geographic interoperability' is the ability of information systems to freely exchange all kinds of spatial information about the Earth and about objects and phenomena on, above, and below the Earth's surface. Secondly, cooperatively, over networks, run software capable of manipulating such information without requiring any knowledge of the underlying infrastructure on the part of the user" (from ISO 19119 – Geographic Services).

The goal of standards and interoperability should be to help GIS users build working systems that are practical to implement and use with the current GIS technology and IT environment. GIS interoperability specifications and standards must fit within the context of broad and widely adopted computing industry standards.

Interoperability can be achieved at several levels. A basic level of interoperability will enable two datasets to be visualized in a common view and it also makes it possible to share geographic data not only with other GIS technologies, but also with non-GIS applications that operate on different platforms. GIS interoperability strategies evolve over time. The enumeration below illustrates this process:

– Export/Import support for Standard Interchange Formats (SDTS, GML, etc.).

– Direct Data Access of Open File Formats (VPF, Shapefiles, etc.).

– Direct Data Access using published Application Programming Interfaces (OGC Simple Features for OLE/COM).

– Common features in a Database Management System (OGC Simple Features for SQL).

– Interaction of standardized GIS Web Services (OGC Web Map Service and Web Feature Service).

Gradually, GIS evolved into georelational structures with limited scalability that enabled related attribute data to be stored in a relational database linked to the file-based spatial features. The dual data structure (i.e. spatial features stored in proprietary file-based format with attributes stored in a relational database) also meant that the GIS could not take full advantage of relational database features such as backup and recovery, replication, and failover. The development of the geodatabase enabled spatial data to be stored in relational databases. Spatially enabled databases opened a new era of broad scalability and made the support of large, non-tiled, continuous data layers possible. When these databases were combined with client development environments that could be embedded within core business applications, the sharing of spatial features with core business applications, such as customer management systems, became a reality.

1.3.4. *Open Geospatial Consortium foundation ideas*

The Open Geospatial Consortium (OGC) promotes a vision in which everyone benefits from geographic information and services made available across any network, application or platform. It is an idea of seamless access to data, independent of how and where it is collected and stored. The OGC has the following recommendation for how geospatial information should be made available:

– Geospatial information should be easy to find and access or acquire, without regard to its physical location.

– Geospatial information from different sources should be easy to integrate, combine or use in spatial analyses, even when sources contain dissimilar types of data or data with disparate feature name schemas.

– Geospatial information from different sources should be easy to register, superimpose and render for display.

– Special displays and visualizations, for specific audiences and purposes, should be easy to generate, even when many sources and types of data are involved.

– It should be easy and without expensive integration efforts to incorporate into enterprise information systems geoprocessing resources from many software and content providers.

1.3.5. *Standardized geospatial Web services*

In the technical respect, SDI encompasses standards, means of data transmission and processing, as well as services, and users can utilize metadata, data and services available in the infrastructure by exploiting utility programs. The concept of services and, more precisely, that of geospatial Web services defined by OGC is key. The aforementioned services enable us to, among other things:

– find information concerning the localization, organization, conditions of use, etc. of data potentially of interest to us (catalog services);

– display this data on a monitor screen, with the capacity to create images visualizing data from various providers, as well as with access to descriptive data for the spatial objects shown in the image (Web Map Service (WMS));

– obtain data of interest to us (Web Feature Service (WFS), Web Coverage Service (WCS)), to the extent that its provision enables further individual processing;

– process the data (coordinate transformation services, image processing services, geospatial analysis services).

The capability to share data has brought much attention to the need for well defined standards in the creation and sharing of metadata for GIS datasets. The ISO has published Metadata standard ISO 19115, which defines metadata content including mandatory, conditional and optional elements, and is applicable to the cataloguing of spatial datasets and clearinghouse activities.

Based on this possibility, WMO has for several years now been carrying out work on a field-specific profile of the ISO 19115 standard which includes elements specific to meteorology and climatology.

The standardization of metadata provides users with an understanding of geographic information and enables the exchange of "data about data". The metadata standards focus on the metadata user communities' need to manage, share and reuse their geographic data and promote global interoperability. Metadata support is crucial to building a strong spatial data infrastructure.

The metadata documents the location of the information, its content and structures, and provides the end-user with detailed information on its appropriate use. Catalog portals make it possible to find data of interest, but can also provide information and tools which enable us to assess, visualize or download them. These capabilities are linked with further services defined by the OGC.

Web mapping services are often used to assist users in geospatial search systems, showing the geographic context and extent of relevant data with respect to reference data. The OGC defines an open interface for the purposes of such services.

A Web Map Server that provides this service can tell other programs maps that it can produce.

The OGC WFS specification has provided a solution for the standardized request and delivery of vector data. Supporting the OGC "Feature Model" it defines the dialog required to interact with geographic features via vector data service.

The OGC Web Coverage Specification extends the Web Map Service (WMS) interface to allow access to geospatial "coverages" that represent values or properties of geographic locations, rather than WMS generated maps (images).

For the capability to utilize the spatial information provided by WMS, WFS and WCS, the format in which it reaches the client is highly important. Consequently, a summary of information about data formats used in geospatial Web services.

Metadata and attribute data are encoded in XML for all of the aforementioned services. As regards geometry data, the format depends on the service type. There is no formal list of required supported formats for WMS. GIF, PNG, JPEG, TIFF, SVG, WebCGM are mentioned here with respect to the necessity to provide at least one format that supports transparency for the portrayal of vector features and at least one format that can be displayed by common Web browsers without additional software. The WFS provides geometric data in Geography Markup Language (GML). Additionally, it is the only server which enables the uploading/updating of data on the server in the case of the so-called transactional WFS. WCS supports at least one of the following formats for each coverage offered: GeoTIFF, HDF-EOS, DTED, NITF, GML. Servers may offer coverages also in other encoding standards.

The geospatial services described above, in practice, define the idea of interoperability on the basis of open standards. From the viewpoint of meteorology and climatology, this shows what the requirements will be for applications to be created in the future in these fields, for the purpose of providing data or services.

1.3.6. *GIS and AS interoperability potential: data model and formats*

Data exchange and/or sharing requires compatibility of the data model and formats. The GIS community, which is represented by, among others, OGC, has proposed several solutions in this area, e.g. GML as a universal language for description of spatial data. There is also the OGC Simple Feature Specification, which is a proposal for the description of objects with respect to geographical space. Therefore, we have NetCDF and HDF which are widely used, especially in the academic community, or BUFR and GRIB, which are oriented towards the exchange of operational data, e.g. results from numerical weather prediction models. It should be remembered that for all of the aforementioned formats, there are libraries of procedures facilitating manipulation and exchange of data, while the last two of these formats are standards for data exchange within the WMO.

1.3.7. *Atmospheric data model*

The ArcGIS Atmospheric Data Model is a collaborative initiative among ESRI, UCAR, NCAR, Raytheon, Unidata and NOAA. The ultimate goal of the ArcGIS Atmospheric Data Model is to represent each of these data objects in a uniform manner, enabling their superposition and combined analysis in the ArcGIS desktop environment. A project is underway for the combined support of NetCDF and HDF-5 through new graphic and animating application tools which will deliver NetCDF converters and a direct-read capability for integrating raster data into the ArcGIS desktop environment. The development of the Atmospheric Data Model provides semantic interoperability between GIS and AS, and can serve two distinct communities: the atmospheric community (which includes atmospheric scientists) and large-scale GIS users.

An Atmospheric SIG with data modeling efforts would establish a working dialog between ESRI and the atmospheric sciences community for ESRI software development.

Two areas need immediate attention, namely temporal data management and improved raster data management with integrated data support. The ability to analyze and manipulate the temporal elements of atmospheric datasets is key to the successful integration into GIS. The challenge with raster data is to make it simple in order to convert new formats to GIS and enable the layering of this data with other GeoSpatial datasets.

The development of NetCDF converters for loading data into ArcGIS is now underway with combined support for NetCDF, HDF and GRIB formats through a single API.

A NetCDF converter that functions in the background would enable the "direct read" of NetCDF data into a GIS, with the option to read satellite data as either points, raster or tables. This will enable a more robust exploration, query and analysis of these data, resulting in a full and complete integration into GIS.

The representation of raster data can be handled in many ways in a GIS. Primarily, raster data that references one "slice" of data in space or time can be represented in the database and thus included in the data model as a single object. Raster data that represents many of these same slices over an extended period of time can be animated to show a high resolution change of phenomenon by using the Raster Catalog as the database storage "Container". The Cell-Value and pixel color representation of raster data is one of the more effective ways to cause leverage on current GIS functionality.

The limited ability of GIS technology to integrate different data types (e.g. biophysical, geophysical, socio-economic, etc.) from different sources, to analyze these data and to present results in an appropriate manner for decision-making, has led to GIS missing data, including but not only atmospheric measurements, weather forecasts and analyses, and climatic data. This data has remained primarily in the hands of the atmospheric science and operational meteorological community, since these data formats are currently incompatible with GIS applications. Isolated commercial data converters have appeared for selected meteorological data in "GIS-ready" formats.

The atmospheric science and GIS user communities would be well served by data interoperability. Spatial data interoperability is a key to interdisciplinary research and decision-making within the atmospheric and GIS communities. The development of an Atmospheric Data Model is one of the first steps towards achieving interoperable datasets.

The Atmospheric Data Model is the first building block in the bridge between the GIS and the atmospheric science communities. "Time is essential for understanding AS (atmospheric science) phenomena. It can be expressed in units ranging from seconds (e.g. rainfall variations measured by a sequence of radar scans) to centuries (climatological variations calculated through complex models). Both running-clock (e.g. experiment time) and epoch-based (e.g. calendar time) approaches are commonly used. For AS data, time and evolution of observed phenomena are as important as spatial location" [NAV 04]. In the GIS environment, time is thought of as an attribute of a feature instead of a dimension of the feature. This significant difference is one issue that the Atmospheric Data Model must resolve. A second challenge is the vertical dimension in atmospheric phenomena. The data model will be a guideline for atmospheric data structures that will facilitate data exchange and interoperability. A successful Atmospheric Data Model will be an effective medium of communication between the atmospheric science and the GIS user communities.

1.3.8. *Support from GIS for atmospheric data formats*

Generally GIS packages do not support atmospheric data formats but there are examples of such support both from commercial and open source solutions.

Feature Manipulation Engine (FME) from Safe Software, apart from support for a gallery of GIS, CAD and database formats, including GML, provides a netCDF reader (beta version).

On the other hand, there is an activity of atmospheric community related to facilitate the interoperability with GIS. The NetCDF Markup Language (NcML) with GIS extension (NcML-GML) could be an example.

1.4. Conclusions

GIS is predominantly used to define stationary phenomena, whereas cartography tends to concentrate on landscape features of a geological timescale or at least seasonal changes such as ground cover, albedo, etc.

The use of GIS for Climatology and Meteorology emphasizes the difference of how data value declines with time and how the temporal resolution adds value to data, as does the integration of prediction, where the accuracy of models can improve the quality of change simulation, mostly for the analysis of Global Change.

Another question relating to data supply, its quality and quantity, the speed of access and data modeling raises concerns about data dissemination with improvements introduced by new technologies.

From the perspective of data analysis processes as the only ones that run at computer speed, problems of data organization and structure, related to Data

Modeling highlight concerns over the representation of spatial-temporal problems and the geo-relational model.

1.5. Bibliography

[CHR 02] CHRISTAKOS G., BOGAERT P. and SERRE M. (2002). Temporal GIS. Advanced functions for field-based applications, Springer-Verlag.

[ESR 05] ESRI (2005). Interoperability in Enterprise GIS, An ESRI White Paper, April 2005.

[ISO 00] ISO/WD 19123.3.3 (2000). Geographic information – Coverage geometry and functions, ISO TC 211/WG2/N138.

[ISO 01] ISO/TC 211/WG 1/WI 19129 (2001). Geographic information – Imagery, gridded and coverage data framework, ISO/TC 211 N 1176.

[ISO 03] ISO/WD 4 19130 (2003). Geographic information – Sensor and data models for imagery and gridded data, ISO TC 211.

[NAV 04] NATIVI S., BLUMENTHAL M.B., CARON J., DOMENICO B., HABERMANN T., HERTZMANN D., HO Y., RASKIN R. and WEBER J. (2004). Differences among the data models used by the Geographic Information Systems and Atmospheric Science communities. Proceedings of the 84th AMS Annual Meeting, 11-15 January 2004, Seattle, USA.

[NAV 05] NATIVI S., CARON J., DAVIS E., DOMENICO B. (2005). Design and implementation of netCDF Markup Language (NcML) and Its GML-based extension (NcML-Gml), Computers & Geosciences, vol. 31, no. 9, November.

OGC specifications (GML, WMS, WFS, WCS, etc.), http://www.opengeospatial.org/specs/?page=specs.

OpenGIS Consortium (1999). The OpenGIS™ Abstract Specification Topic 6: The Coverage Type and its Subtypes Version 4.

[PET 01] PETROSYAN A.S. (2001). GIS in meteorology and climatology. The needs and the challenges. European Geophysical Society, XXVI General Assembly, Nice, 25-30 March 2001.

[SHI 96] SHIPLEY S.T., GRAFFMAN I.A. and BEDDOE D.P. (1996). GIS does the weather. Proceedings, ESRI User Conference, 20-24 May 1996, Palm Springs, USA.

[SZN 05] SZNAIDER R.J. (2005). Operational Uses of Weather Information in GIS-based Decision Support Systems. (http://www.meteorlogix.com/pdf/MxInsight/METEORLOGIX_MXI_WHTPAPER.pdf).

Spatial Data Infrastructure Cookbook, http://www.gsdi.org/gsdicookbookindex.asp.

[WIL 05] WILHELMI O., BETANCOURT T., BOEHNERT J., SHIPLEY S., BREMAN J. (2005). ArcGIS Atmospheric Data Model (Draft).

Chapter 2

SIGMA: A Web-based GIS for Environmental Applications

2.1. Introduction

The large amount of climatological and meteorological information produced by many countries during the last decades facilitated the creation of a huge and varied database. Meteorological agencies spread all over the world yield lots of information about atmospheric conditions and their interactions with oceanic and continental surfaces. The different types of data produced by numerical models, satellites, telemetric stations and others present different formats and bring information which reproduces the weather conditions in different areas and in different temporal and spatial resolutions.

The information produced by numerical models and also by satellite displays values in fields which can be represented by vectors (isolines) or grid points (pixels). Thus, they represent variables whose main characteristic is their spatial continuity. Variables measured by telemetric stations display punctual values; however, they can become continue in space when interpolated by a mathematical method.

All of this data has a strong relationship with the physical space and, in order to be analyzed in a context which involves human and natural aspects, this information requires advanced tools to aid its processing when searching for inferences from the processing of data with different nature and formats. In this sense, Geographical

Chapter written by Carlos Frederico ANGELIS, Fabiano MORELLI, Luiz Augusto TOLEDO MACHADO and Cintia Pereira DE FREITAS.

Information Systems (GIS) assume an important role in meteorology and climatology because it is one of the most powerful and efficient tools to manipulate spatial data.

According to [CAM 96] a GIS not only makes it possible to capture, model, handle and retrieve, but also to analyze georeferenced data by using dedicated hardware and software. The GIS can create and display maps and can also carry out tasks used in the planning and management of different cases. Among all applications, the GIS is largely employed in agriculture, environment, forestry, meteorology, networks monitoring services, etc. [LON 01].

The majority of commercial GIS has functionalities which make it possible to process and analyze spatial data, however, in some cases those functionalities are not enough and it is necessary to develop specific functions and procedures. In this way, the customization of GIS is a fact and has become common in recent years. Moreover, the feasibility of making a GIS available on the Internet, where users can use it instead of installing it on their desktop increases the power of this technology.

The complexity of the information hosted in a GIS requires standardization for different data patterns in order to increase their interoperability and/or data exchange. This issue has been discussed by different communities which deal with georeferenced data. Some patterns have been proposed by the OpenGIS (Open Geographic Interoperability Specification) [BUE 98, GAR 96]. The suggestions of the OpenGIS are disseminated by the Open Geospatial Consortium (OGC; http://www.opengeospatial.org), which is a non-profit, international, voluntary consensus standards organization that is leading the development of standards for geospatial and location based services. Through the OGC's member-driven consensus programs, OGC works with the government, private industry and academia to create open and extensible software application programming interfaces for GIS and other mainstream technologies.

2.2. CPTEC-INPE

The "Centro de Previsao de Tempo e Estudos Climaticos" of the "Instituto Nacional de Pesquisas Espacias" (CPETC-INPE) produces a large amount of data in different temporal and spatial resolutions which is used for different purposes. CPTEC is an operational weather forecast center and the data produced by its internal departments is made available to users mainly on the Internet.

Considering that the Internet is the main communication channel between users and CPTEC, it is necessary to offer users an efficient interface which facilitates the search for information and gives them extra tools to integrate, visualize and analyze data of different sources and patterns.

Bearing that in mind, CPTEC has developed an interface of integration and visualization of products by using Mapserver technology which is in accordance with the recommendations suggested by the OpenGIS. According to the official MapServer website (http://mapserver.gis.umn.edu) it is an open source development environment for building spatially-enabled Internet applications. MapServer excels at rendering spatial data (maps, images and vector data) for the Web. Although MapServer is not a full-featured GIS system, it meets the requirements needed to become one. Depending on the needs of the user, it is possible to develop tools and applications dedicated to GIS applications by using embedded programs.

The use of Mapserver technology to create the GIS hosted at the INPE/CPTEC Website is the issue with which this chapter deals. This GIS is named SIGMA (GIS for Environmental Applications) and is described in the next sections.

2.3. SIGMA

For an ordinary user who searches meteorological information to plan his daily activities, the use of a GIS is much more than a simple "look at a single map" which usually shows static and isolated information about a specific theme. By using a GIS it is possible to combine a lot of information and visualize the result in form of layers which contain different data. The integration of data is one of the primary functionalities of a GIS.

SIGMA in its full conception will be a proper Web-based GIS developed to attend the needs of CPTEC's users not only for basic consultation, but also for advanced operations which use special tools designed to obtain new information. The diagram showing the main functionalities of SIGMA is illustrated in Figure 2.1.

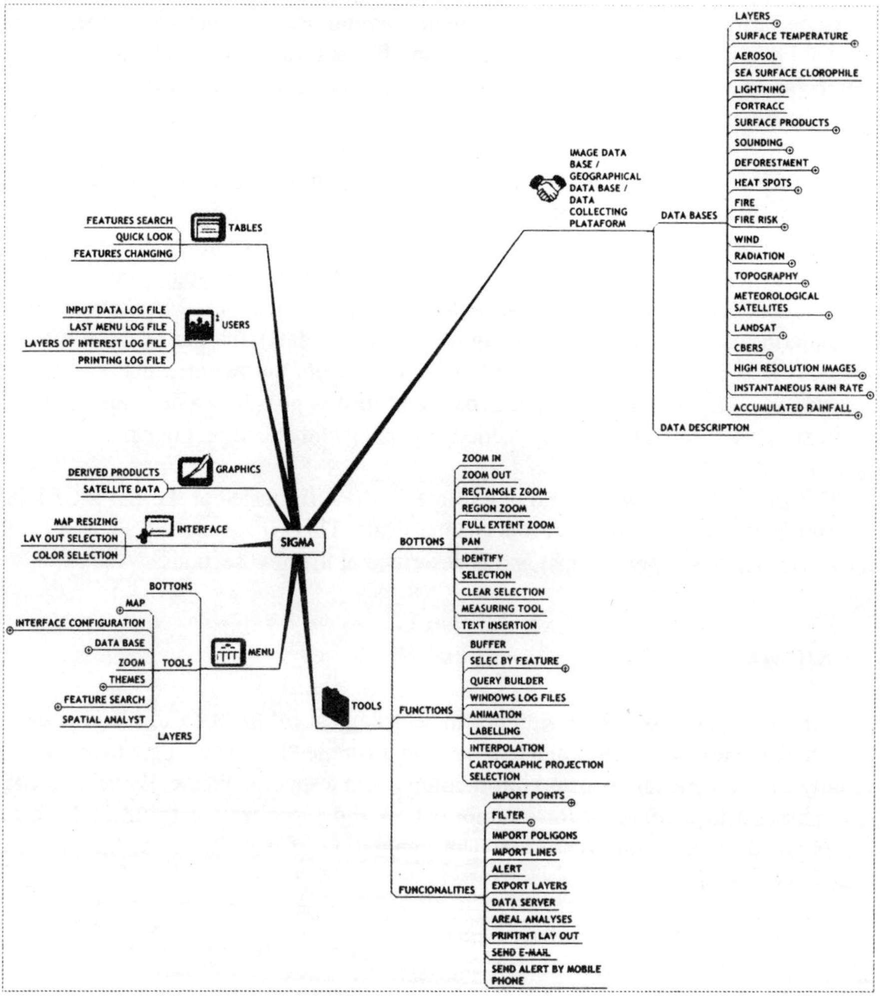

Figure 2.1. *SIGMA diagram and its main functionalities*

SIGMA provides users with resources to visualize, consult and carry out spatial analysis of data which is generated by CPTEC and also enables users to upload external data to be processed together with SIGMA's own data. According to Figure 2.1, users can handle information with spatial and tabular features, monitor weather and environmental conditions in real-time, produce reports and search for historical information in the CPTEC database.

2.3.1. *Basic functions*

The system presents basic functions like zoom in and zoom out, query builder, distance measuring, overlay of multilayers, upload of punctual, etc. Within the SIGMA environment the user can find pre-existent layers like topography, rivers, roads, cities, county and country borders, rail tracks, airports location, LANDSAT/TM mosaic, etc. All of those extra layers cover the entire South America and make SIGMA suitable for users in the whole continent. Operational meteorological products like lightning occurrence, rainfall estimated by satellite and radar, ultraviolet indices, amount of ozone in the atmosphere, burning occurrence, sea surface and continental temperature, etc. which are generated by the CPTEC are also available within the SIGMA.

2.4. Impacts of weather conditions on the economy

The economies of many countries and the way of life of their societies are directly affected by weather conditions. Not only the simple act of going out of our home but also complex tasks like the planning and management of all logistics involved in aerial or terrestrial transportation of people and goods, for instance, can be influenced by weather conditions. Therefore, the monitoring of weather conditions can be beneficial to several areas.

The high demand for information which enables the monitoring of weather conditions requires a special system which integrates the functions of GIS with the operational methods of data generation. This type of system can help decision makers who need an automatic system to monitors weather conditions in order to emit alerts and/or reports when an extreme weather event occurs. This type of system is, nowadays, a new challenge in terms of environmental monitoring.

2.5. Severe Weather Observation System (SOS)

The nowcasting products generated by CPTEC from the detection of lightning occurrence, radar and satellite data are stored within SIGMA in order to improve the monitoring of weather conditions over some parts of South America. An automatic system makes use of that data and evaluates the risk of storms, hail, lightning and flash flood occurrence over a given area. This system is named SOS and is embedded in SIGMA.

2.5.1. *Tracking of convective clouds*

SOS monitors convective clouds and tracks their origin, expansion rate, life cycle and both temporal and spatial evolution, as well as lightning occurrence associated with each convective cell. According to Machado and Laurent [MAC 04] the area of a convective cloud can be estimated as a function of its life duration (in hours) and is expressed by:

$$A(t) = \alpha * e^{at^2 + bt + c}$$

where α, a, b and c are parameters extracted from the total life cycle of the convective system. The tracking of a convective system enables estimates of its position in the future and also its propagation speed by a linear extrapolation based on its previous position and speed.

2.5.2. *Risk of lightning occurrence*

According to [SCH 97] and [KUR 97] the positive difference between thermal infrared (10.5 μm) and water vapor (6.7 μm) GOES channels is possibly related to deep convection, once under these circumstances the updrafts can inject water vapor in the high levels of the atmosphere, usually reaching the stratosphere. According to Figure 2.2 when this occurs the brightness temperature of the water vapor channel tends to be greater than the thermal infrared of the cloud tops which are located in upper troposphere.

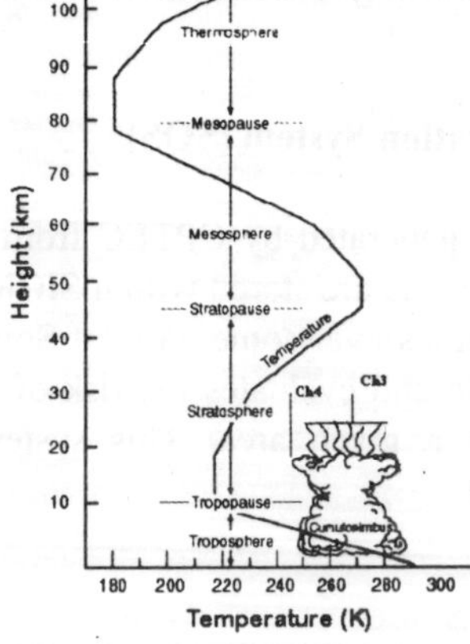

Figure 2.2. *Vertical profile of temperature according to the altitude*

Figure 2.2 shows that above the tropopause there is an inversion in the vertical profile of the temperature which causes the high values of water vapor brightness temperature. Convective clouds which present these characteristics usually produce lots of lightning and it is possible to relate the number of lightning occurrence with the difference between channel 3 and channel 4 of the GOES-12 satellite. According to [MAC 05] the relationship between this difference and the amount of lightning occurrence on the surface is shown in Figure 2.3.

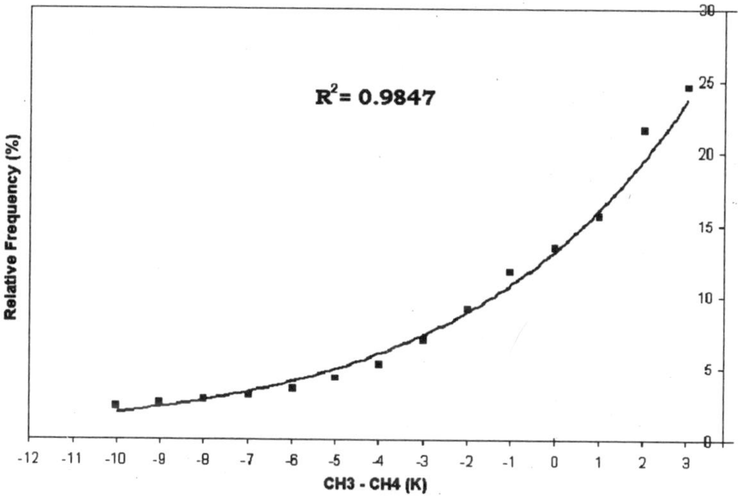

Figure 2.3. *Probability of lightning occurrence. The values shown in the graph represent the frequency of lightning observed on the surface during two consecutive GOES-12 images as a function of the difference of brightness temperature between channel 3 and channel 4*

2.6. SOS interface

Figure 2.4 exhibits the SOS interface which appears in the Web browser when the user logs on the site. Similar to what happens in SIGMA, users can choose which layer and/or products they want to visualize and also see if there is an alert emitted. When a critical situation is registered, for example, if a storm is predicted to occur in one hour's time, the regional borders where this storm will possibly occur will be highlighted in the screen (see Figure 2.4). In this case, SOS can send a message to someone by email and also by mobile phone.

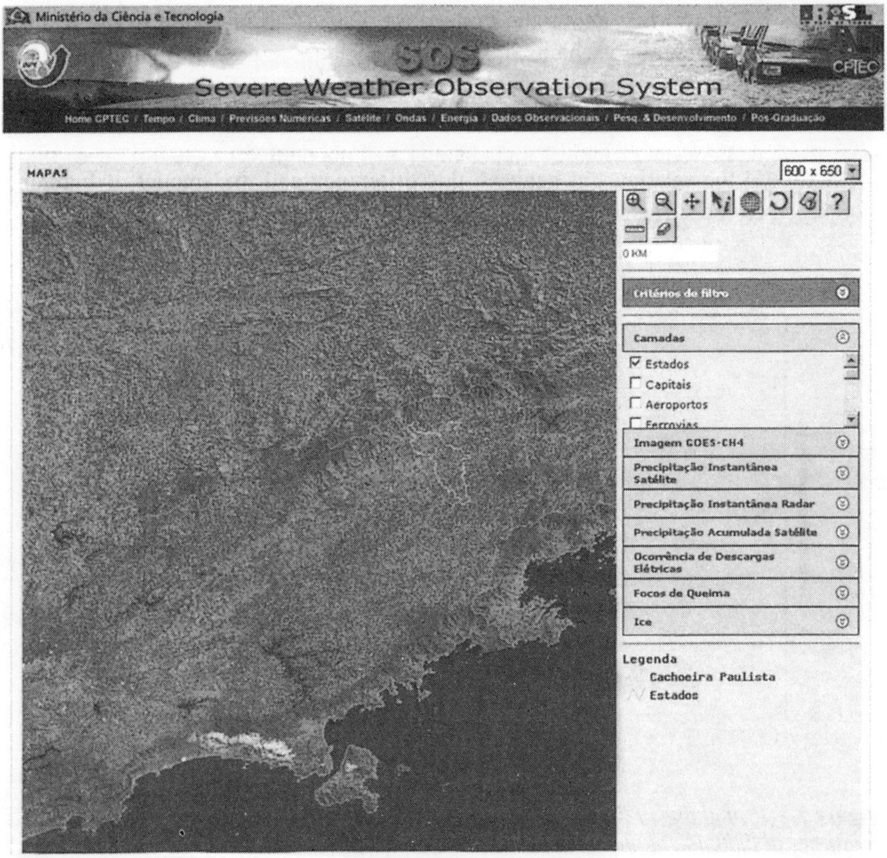

Figure 2.4. *SOS interface for Web browser*

From the SOS interface the user can see the past evolution of an extreme event and also its future evolution which is assessed by the nowcasting techniques. As the SOS is embedded in SIGMA, all visualizations can be integrated with other layers or products which are available in SIGMA. Figure 2.5 illustrates a storm predicted by the tracking of convective clouds and shows the estimated rain rate associated with the storm.

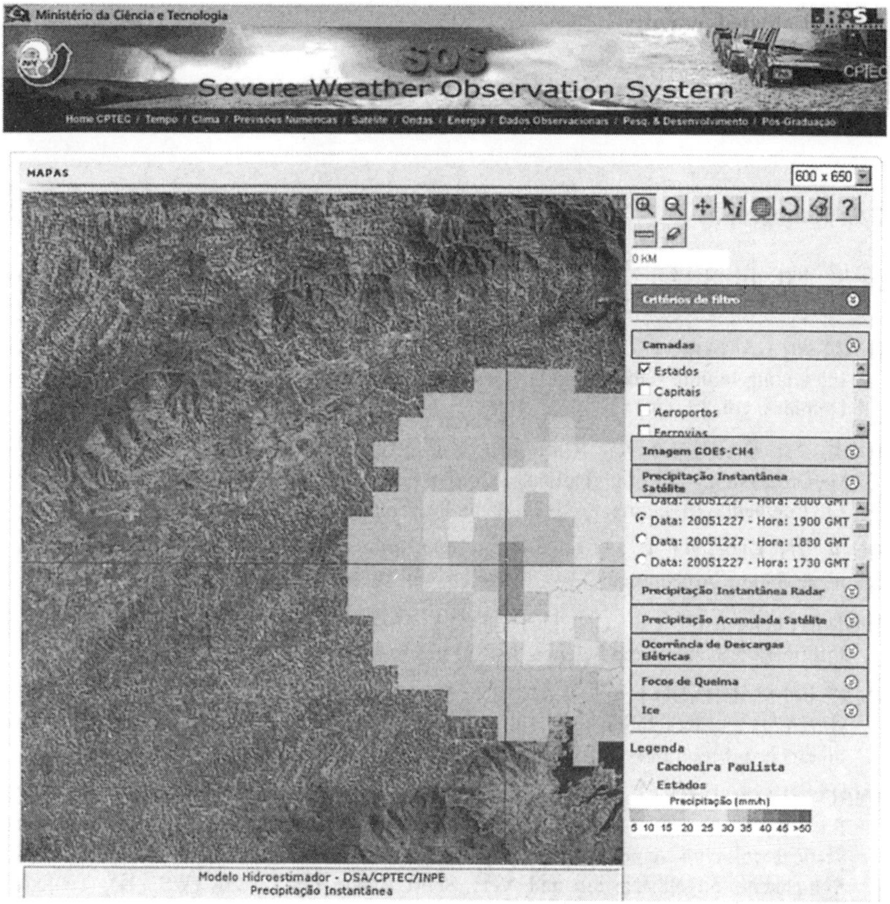

Figure 2.5. *Storm predicted by nowcasting techniques and emission of an alert by the SOS*

2.7. Conclusions

Weather conditions can have an impact on society as a whole and are sometimes responsible for lives and economical losses. Minimizing these losses is one of the main challenges for meteorologists and climatologists because this task requires efficient automatic systems which can predict the occurrence of extreme weather events reliably. Automatic systems developed to alert about bad weather conditions must be built to help decision makers improve their capacity to let authorities know when and where an extreme event will occur. Systems like SOS and SIGMA showed above are examples of the use of GIS technology to offer users advanced tools to aid them in dealing with different types of information at the same time.

2.8. Acknowledgements

This work was supported by Fundacao de Amparo a Pesquisa do Estado de Sao Paulo – FAPESP (Proc: 03/10508-0).

2.9. Bibliography

[BUE 98] BUEHLER K., The OpenGIS™Guide – Introduction to Interoperable Geoprocessing. Ed. K. Buehler and L. McKee, Massachusetts, 1998.

[CAM 96] CAMARA G., SOUZA R.C.M., FREITAS U.M., GARRIDO J.C.P., SPRING: Integrating remote sensing and GIS with object-oriented data modelling. Computers and Graphics, vol. 15, no. 6, p. 13-22, 1996.

[GAR 96] GARDELS K., The Open GIS Approach to Distributed Geodata and Geoprocessing. 3rd International Conference/Workshop on Integrating GIS and Environmental Modeling, p. 21-25, Santa Fé, January 1996.

[KUR 97] KURINO T., A satellite infrared technique for estimating "deep/shallow" precipitation, Advances in Space Research, vol. 19, Issue 3, p. 511-514, 1997.

[LON 01] LONGLEY M.F., GOODCHILD D.J., MAGUIRE D.J., RHIND D.W., Geographic Information Systems and Science, John Wiley and Sons, 2001.

[MAC 04] MACHADO L.A.T., LAURENT H., The convective system area expansion over Amazonia and its relationships with convective system life duration and high-level wind divergence. Monthly Weather Review, vol. 132, no. 4, p. 714-725, 2004.

[MAC 05] MACHADO L.A.T., MORALES C.A., LAURENT H., VILA D., RODRIGUES S.M., LIMA W.F.A., Convective system area expansion, high-level wind divergence and vertical velocity: A tool for nowcasting, in World Weather Research Programme's Symposium on Nowcasting and Very Short Range Forecasting (WSN05), Toulouse. 2005.

[SCH 97] SCHMETZ J., TJEMKES, S.A., GUBE, M., VAN DE BERG L., Monitoring deep convection and convective overshooting with METEOSAT, Advances in Space Research, vol. 19, Issue 3, p. 433-441, 1997.

Chapter 3

Web Mapping: Different Solutions using GIS

3.1. Introduction

The Internet can be used in an effective manner to visualize as well as provide access to information for a wide range of users. This also applies to presentation and provision of access to numerical maps (Web mapping). This latter task is often performed on the basis of GIS technology, since the GIS packages available on the market offer Internet map servers as a standard software element. Open-source solutions can also be used. In both cases, we have at our disposal solutions which provide unified access to spatial information based on OGC standards[1] and fulfill modern interoperability requirements[2].

Aside from the direct usage of GIS for Web mapping, the usage in offline mode is also possible. In this case, GIS tools are used to prepare a Web map service which will later be made available to the user without the involvement of GIS tools. Here as well, we can use inexpensive or free tools, which are normally compatible with desktop GIS packages, in order to prepare dedicated solutions which do not require the aforementioned spatial information access standards to be fulfilled.

Chapter written by Pawel MADEJ, Malgorzata BARSZCZYNSKA and Danuta KUBACKA.

1 The Open Geospatial Consortium (http://www.opengeospatial.org) is an international non-profit organization whose aim is to standardize geospatial services.

2 According to the definition provided by the US Institute of Electrical and Electronics Engineers, *interoperability* is the ability of two or more systems or components to exchange information and to use the information that has been exchanged.

Below, we present our experiences concerning the building and usage of Internet services providing access to spatial information from areas of meteorology and climatology. We shall present several examples which use GIS tools in both offline and online mode.

3.2. Examples of Web mapping based on the usage of GIS technology in offline mode

The preparation of Internet maps based on the usage of GIS technology in offline mode requires the following steps to be taken:

– definition of the content and graphic form of the map: the range of objects and attribute data, as well as symbols representing map objects with the aid of the chosen GIS software;

– automatic generation of a set of dHTML/XML pages and/or database content for the Web service, with the aid of appropriate tools (script or GIS package extension).

The experiences described below concern solutions which are compatible with the ESRI software (ArcView). The examples are based in large measure on "open-source" tools and, on the user hand, all that is needed is a Web browser. Depending on needs, it is possible to prepare ready-made Web pages or dynamically create them on the Web server side.

In the simplest case, it is possible to prepare a set of map data presented on the Internet by using raster images containing pre-selected elements of a ready-made map. The information source is an ArcView project file, which is the basis for generating a set of dHTML files with the aid of an ArcView extension (e.g. HTML ImageMapper, http://www.alta4.com). Aside from the map image itself, the user obtains access to descriptive information concerning the indicated map element. The prepared maps can have several levels of detail: the less precise maps serve the purpose of assisting in the selection of the area of interest to the user, whereas the more detailed maps provide, aside from graphic information, access to descriptive information (see [BAR 00]). A disadvantage of this solution is the finished, predefined set of maps which is available to the user and the advantages are the possibility of adapting visual information for presentation purposes, as well as the simplicity of use. Also important is the fact that, in the solution being discussed, the graphic form of the map can be prepared in a GIS environment, which enables its wealth of capabilities to be used. In a similar manner, it is possible to prepare a service providing access to vector images, with the option of adding a raster background. The advantage of this solution is the continuous scaling of maps and the possibility of selecting the range of information to be shown on the map. In both cases, the maps prepared can be made available over a network, or on CD-ROM as an independent set of dHTML pages.

Another solution is to convert the GIS data which is of interest to us into graphic formats and place it in a dedicated database for the mapping service. In this solution, maps are generated on demand, as in traditional Web map servers; "in-flight" conversion is no longer necessary. For the presentation of maps we can use Scalable Vector Graphics (SVG), which is a relatively new language from the XML family, oriented towards the description of two-dimensional graphics and recommended by the W3 Consortium (see [NEU 03]). Many examples of projects and tools solutions based on SVG can be found at the following address: www.carto.net. Solutions available here, e.g. OpenSVGMapServer, have been used in the examples presented below.

We shall show the possibilities for the usage of GIS technology to generate services presenting maps on the Internet with an example of a solution enabling the presentation of information about the IMGW measurement network, which is prepared with the aid of ArcView and free scripts. This example presents basic information about the IMGW measurement network in the San River basin (an area of c. 17,000 km^2 located in south-eastern Poland and Ukraine), as well as about the possibility of obtaining archival data (hydrological and climatological). It is an interactive vector map providing the possibility of selection of active layers, continuous scaling and panning of a map. The objects on the map respond to the movement of the cursor, as well as to a click of the mouse, and in this way the user can obtain information about the measurement stations of interest to him (attribute data). The content of the service is a sort of meta-base about available measurement information.

Two solution variants were prepared. In both variants, the service's code was written largely automatically, on the basis of data from GIS database information layers collected in the form of an ArcView project file. The first variant uses HTML and SVG technology for map presentation, as well as Javascript for user interaction. The service can be disseminated via a Web server or offline, e.g. on CD-ROM. The second variant possesses similar functional characteristics, but operates on the basis of a database containing a GIS information layer content, which is converted into graphic formats. This enables the making of simple queries, as well as object filtering. The technologies used are – as before – HTML, SVG and Javascript. Additionally, it is necessary to use a Web server (e.g. Apache), database (MySQL) and access language (PHP). The dHTML/SVG files are generated dynamically at the Web server end. PHP scripts serve the purpose of preparing information corresponding to the user's query, which significantly reduces the volume of data transmitted from server to client and therefore shortens the waiting time. The information service described is presented in Figure 3.1. Further examples can be found in [MAD 02].

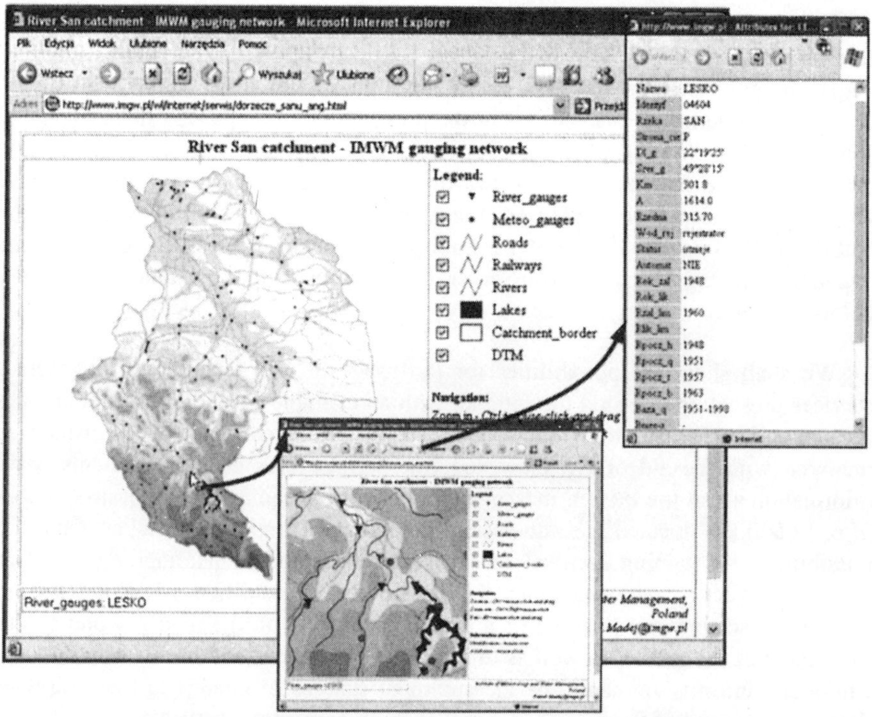

Figure 3.1. *Interactive maps presenting the Institute of Meteorology and Water Management measurement network (Poland)*

The Internet map access capabilities and examples presented above are oriented towards solutions serving concrete purposes for a human client with well-defined needs. Present trends in the area of spatial information access emphasize the widest possible information access, the capability of integrating data from various sources and the standardization of geospatial data and services. Solutions fulfilling these postulates will be presented in the next section.

3.3. Examples of Web mapping using GIS tools in online mode

The modern notion of geo-information use is closely connected with the widespread (on-demand) access to data. It assumes that the institutions providing data are willing to make it available. At the same time, "beneficiary" institutions should have the possibility and tools to enable effective data browsing and access to widespread data. A concept of spatial data infrastructure (SDI) has even appeared whose aim is to emphasize that an information infrastructure is just as important for the development of societies as a technological infrastructure. At the worldwide and European level, we observe operations aiming in the direction of building an SDI on as broad a scale as possible – for example, the activity of the non-commercial organization Global Spatial Data Infrastructure (http://www.gsdi.org), or the INSPIRE initiative (http://inspire.jrc.it), as part of which the project for a directive regulating these issues in the European Union was begun.

In the technical respect, SDI encompasses standards, means of data transmission and processing, as well as services, and users can use metadata, data and services which are available in the infrastructure, by making use of utility programs. The concept of services is key – more precisely, that of geospatial Web services defined by OGC. The implementation specification for a Web map service, which was prepared by the OGC has recently become an ISO standard[3]. Next, specifications concerning dissemination of spatial data or catalog services, enabling the search of geospatial data and services accessible on the Web, which are important for the problem under consideration, will soon become ISO standards, but already are *de facto* norms. The aforementioned services enable us to, among other things (see [GSDI 04]):

– find information concerning the localization, organization, conditions of use, etc. of data potentially of interest to us (catalog services);

– display this data on a monitor screen, with the capacity to create images visualizing data from various providers, as well as with access to descriptive data for the spatial objects shown in the image (Web map service (WMS));

– obtain data of interest to us (Web feature service (WFS), Web coverage service (WCS)), to the extent that its provider has made it available to us for further processing for our own purposes;

– process the data (coordinate transformation services, image processing services, geospatial analysis services).

The description above illustrates the idea of on-demand access to data and services, according to which providers deliver to users, in online mode, information and data according dynamic queries which are formulated by the users. Below, we will describe WMS and WFS in more detail, as well as present examples of their usage for the presentation and dissemination of results from spatial analyses of climatological and meteorological fields.

3 The ISO 19128 standard entitled "IS 19128:2005 Geographic information – Web map server interface".

The WMS is probably the most widespread geospatial service, and the majority of GIS software providers have WMSs which are compliant with the aforementioned specifications in their product line. There are also many specialized solutions, including freeware tools as well, such as Mapserver, which is a product built at the University of Minnesota and developed as part of the UMN Mapserver Project (http://mapserver.gis.umn.edu). A WMS which provides this service can: tell other programs what maps it can produce, produce a map (generally as an image) and answer basic queries about the content of the map, i.e. provide attribute data. This enables the service client to obtain a spatially-oriented map in one of the popular graphic formats. The user can compose the map content himself, select its spatial range and obtain information about objects which are located on the map, if the service provider makes such information available. A service client, e.g. standard Web browser, can ask a map server to do these things just by submitting requests in the form of URLs using Key-Value syntax. The content of such URLs depends on which of the three tasks is requested. An alternative form of communication with a map server is XML-based messages. In the same way, one can communicate with a Web Feature Server. The OGC Web Feature Service Specification provides a solution for the standardized request and delivery of vector data.

A Web map server software from GIS packages is normally quite extensively developed and encompasses, aside from the map server software itself, tools for the creation of Web map services, as well as tools and templates for the creation of client applications[4]. Thus, the users' most frequent contact with a map server is contact via a Web browser connecting with the so-called "service browser", i.e. client application. The examples presented are based on ESRI's ArcIMS v. 9.1, working in a client-server environment and compliant with OpenGIS WMS and WFS specifications. The service browsers prepared for the user enable the visualization of the geospatial and attribute data in the client Web browser. The ArcIMS also enables the use of products from the ESRI family (e.g. ArcMap), or other applications which are compliant with OpenGIS standards, on the client side. Therefore, it is also possible to use the information provided by the Web map server in a client map service, use the image from the map server in a user/client information bulletin, as well as for the vector services (WFS) use data contained in the service in further geospatial analysis. The idea of software elements which are necessary for the realization of WMSs, in the example of ArcIMS, is shown by Figure 3.2.

4 Similar capabilities are offered by open-source solutions. An example could be the aforementioned UMN Mapserver which is supplemented with, for example, MapLab (see http://www.maptools.org/maplab) as a tool for the creation of WMSs.

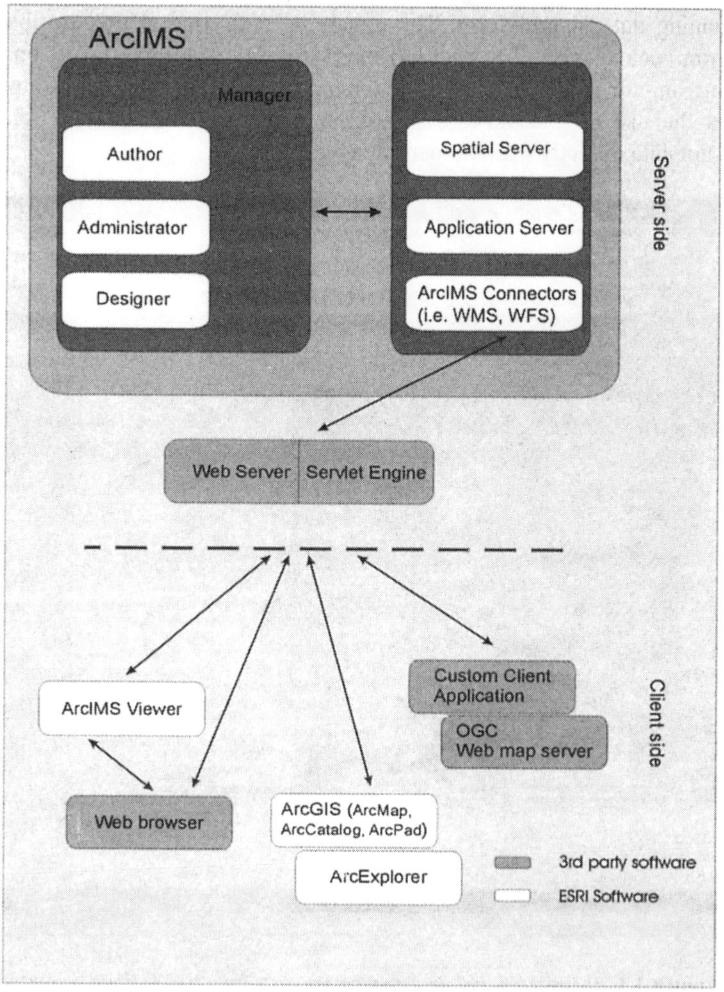

Figure 3.2. *Diagram illustrating the software elements which are needed to build a Web map service using ArcIMS*

Below, we show examples using the technologies employed to enable a more widespread access to results of climatological and meteorological analyses which were described above, as well as using these results in further spatial analyses.

The first example concerns a Web service visualizing the average annual air temperature distribution in Poland (see Figure 3.3). The data source in this case was a layer of measurement data points which was subjected to a so-called spatialization, thus obtaining data in grid form. The remaining data used to prepare the service comes from vector layers. The user has the possibility of turning layers on and off, navigating on the map, as well as zooming in and out. The service presented illustrates the use of the WMS standard, in which only images are sent to the Internet, not data directly from numerical layers.

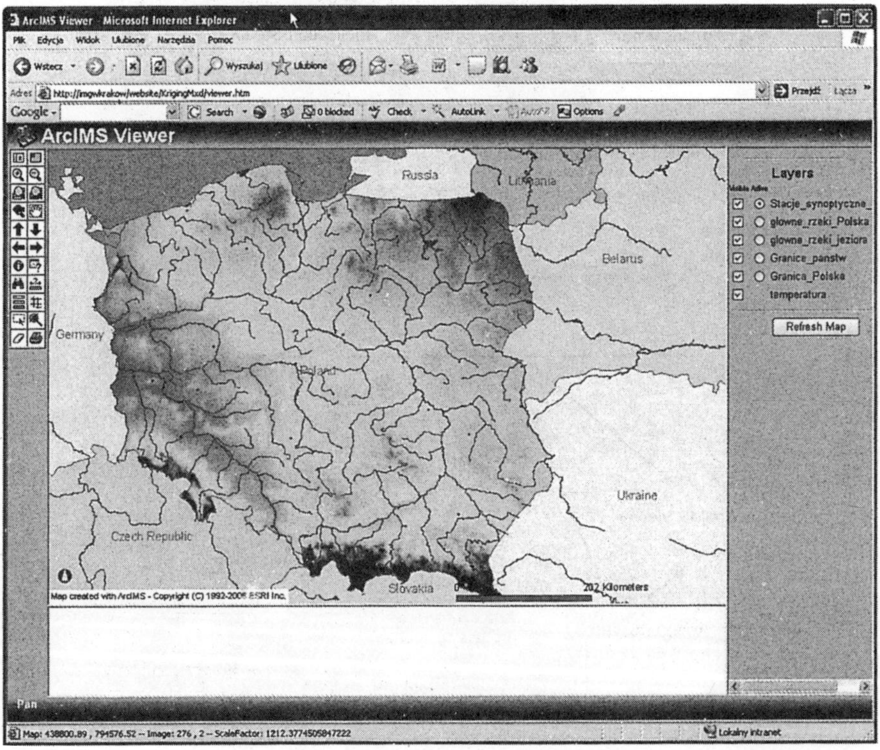

Figure 3.3. *Average annual air temperature distribution in Poland: example using WMS to disseminate the results of climatological analyses*

The second example shows the possibilities for the use of WFS. In the technology presented above, a service was built which provides access to data from the interpretation of a satellite image, which assesses the potential amount of water in clouds. This data was then used by locally-operating GIS software (ArcMap) with the aim of linking it to catchment area data. The effect was a spatial analysis resulting in the calculation of the average potential sum of precipitation for individual catchment areas, as well as in the visualization of these averages (see Figure 3.4).

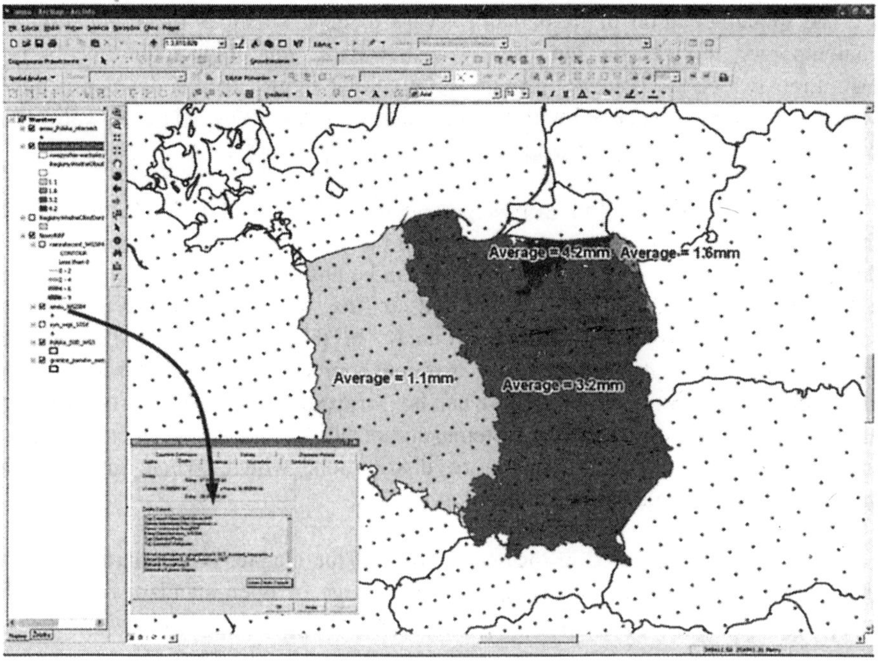

Figure 3.4. *Average precipitation in Polish catchment areas, assessed on the basis of satellite image interpretation: example using WFS as a data source for further analyses*

In the above examples, commercial software was used. Similar effects can be achieved using open source tools (see [SON 04]).

3.4. Conclusion

The idea of widespread access to data and information having spatial references and, in particular, to data on the environment is becoming more and more popular. Its implementation requires the standardization of means for information exchange and delivery of geospatial services. This is the purpose of actions aiming to prepare open standards in this area, as well as of legal and organizational solutions such as the INSPIRE directive. What is becoming more and more universal is the building of so-called geoportals[5], which, by using the most recent achievements in the area of informatics and telecommunications, make spatial information in a broad sense universally available via the Internet.

5 Geoportals are e-commerce sites that enable browsing, viewing and procuring of spatial databases and orthophotography online.

The implementation of on-demand access to data, which is the foundation of contemporary thought on the use of spatial data, sets high requirements for institutions which are potential providers. Of capital significance is development of services which are associated with the provision of access to metadata, as well as to data which is based on the network standards of geospatial services.

An advantage of the solutions based on open standards, such as WMS or WFS, is the open specification of communication with the server providing access to spatial data. This enables the service client, which can be any utility software in which the aforementioned standards have been used, to use the data made available via Internet/intranet to prepare its own products. In this manner, it is possible to integrate and process data coming from different sources. In the case of issues which are associated with meteorology, these can be, for example, data from ground-based and remote sensing measurement systems, and results of analyses such as a satellite image interpretation or a temperature distribution, which, when linked with geographic data, enable various types of usage.

The examples presented in the second part of the chapter are in harmony with these ideas and represent an example for the usage of open standard-based Web mapping in issues which are associated with meteorology and climatology.

3.5. Bibliography

[BAR 00] BARSZCZYŃSKA M., MADEJ P., "Wykorzystanie GIS do przygotowania serwisów internetowych opartych o mapy, na przykładzie serwisu o sieci pomiarowej IMGW", mat. 11 Ogólnopolskiej szkoły gospodarki wodnej pt." Systemy informacyjne do sterowania zasobami wodnymi w warunkach ekstremalnych" Czorsztyn 11-13.11.200, KGW PAN, IMGW, 2000, str. 7-18 (in Polish).

[GSDI 04] GSDI (editor: Douglas D. Nebert), Developing Spatial Data Infrastructures: The SDI Cookbook v.2, January 2004 (http://www.gsdi.org/gsdicookbookindex.asp).

[MAD 02] MADEJ P., "Map-based Internet information services - examples of GIS utilization in off-line mode", Environmental Communication in the Information Society, W. Pillman, K. Tochtermann (eds.), Proc. of 16th Conference "Environmental Informatics 2002", Vienna, 25-27 September, 2002, vol. 2, p. 325-328.

[NEU 03] NEUMANN A., WINTER A., "Vector-based Web Cartography: Enabler SVG", version 2.01, 2003 (http://www.carto.net/papers/svg/index_e.shtml).

[SON 04] SONG X., KONO Y. and SHIBAYAMA M., The development of Web mapping application using open source GIS solution. Proc. of the International Symposium on GeoInformatics for Spatial-Infrastructure Development in Earth & Allied Sciences (GIS-IDEAS), Hanoi, Vietnam, 16-18 September 2004 (http://gisws.media.osaka-cu.ac.jp/gisideas04/viewpaper.php?id=10).

Chapter 4

Comparison of Geostatistical and Meteorological Interpolation Methods (What is What?)

4.1. Introduction

Firstly let us consider the abstract diagram of meteorological examinations. The initial stage is meteorology that means the qualitative formulation of the given problem. The next stage is mathematical calculation in order to formulate the problem quantitatively and the third stage is to develop a software on the basis of the mathematical calculations. Finally, the last stage is again meteorology, that is, the application of the developed software and the evaluation of the obtained results. In practice, however, the mathematical calculations are sometimes neglected. Instead of an adequate mathematical formulation of the meteorological problem, ready-made software is applied to solve the problem. Of course in this case the results are not authentic either. Let us quote John von Neumann: "without quantitative formulation of the meteorological questions we are not able to answer the simplest qualitative questions either".

Concerning our topic, we have the following question: what kind of spatial interpolation mathematics is adequate for meteorology? Nowadays geostatistical interpolation methods built in GIS software are applied in meteorology. The mathematical basis of these methods is geostatistics, which is an exact but special part of mathematical statistics. The specialty is connected with the assumption that

Chapter written by Tamás SZENTIMREY, Zita BIHARI and Sándor SZALAI.

the data is purely spatial. Consequently, the geostatistical methods cannot efficiently use the meteorological data series whereas the data series make it possible to obtain the necessary climate information for the interpolation in meteorology.

4.2. Mathematical statistical model of spatial interpolation

In practice many kinds of interpolation methods exist and therefore the question is what is the difference between them? According to the interpolation problem the unknown predictand $Z(\mathbf{s}_0,t)$ is estimated by use of the known predictors $Z(\mathbf{s}_i,t)$ ($i=1,...,M$), where the location vectors $\mathbf{s}$ are the elements of the given space domain D and t is the time. The vector form of the predictors is $\mathbf{Z}^{\mathrm{T}}(t)=[Z(\mathbf{s}_1,t),.....,Z(\mathbf{s}_M,t)]$. The type of adequate interpolation formula depends on the probability distribution of the meteorological element in question. In this chapter only the linear or additive formula is detailed, which is appropriate in case of normal distribution.

4.2.1. *Statistical parameters*

In general, the interpolation formulae have some unknown interpolation parameters which are known functions of certain statistical parameters. At the linear interpolation formulae the basic statistical parameters can be divided into two groups known as deterministic and stochastic parameters.

The deterministic or local parameters are the expected values $\mathrm{E}(Z(\mathbf{s}_i,t))$ ($i=0,...,M$) and let $\mathrm{E}(\mathbf{Z}(t))$ denote the vector of expected values of predictors, i.e. $\mathrm{E}(\mathbf{Z}(t))^{\mathrm{T}}=[\mathrm{E}(Z(\mathbf{s}_1,t)),....,\mathrm{E}(Z(\mathbf{s}_M,t))]$.

The stochastic parameters are the covariance or the variogram values belonging to the predictand and predictors such that:

- $\mathbf{c}$: predictand-predictors covariance vector;
- $\mathbf{C}$: predictors-predictors covariance matrix;
- $\boldsymbol{\gamma}$: predictand-predictors variogram vector;
- $\boldsymbol{\Gamma}$: predictors-predictors variogram matrix.

The covariance is preferred in mathematical statistics and meteorology, whereas the variogram is preferred in geostatistics. Here is a quotation from the chapter "Geostatistics" of ([CRE 91], p. 30): "the cornerstone is the variogram, which is a parameter that in the past has been either unknown or unfashionable among statisticians". The main reason for this reluctance is that the covariance is a more general statistical parameter than the variogram. The variogram values can be written as functions of the covariance values but the opposite is not true.

4.2.2. *Linear meteorological model for expected values*

At the statistical modeling of the meteorological elements we have to assume that the expected values of the variables are changing in space and in time alike. The spatial change means that the climate is different in the regions. The temporal change is the result of the possible global climate change. Consequently in the case of linear modeling of expected values we assume that

$$\mathrm{E}(Z(\mathbf{s}_i,t)) = \mu(t) + E(\mathbf{s}_i) \quad (i = 0,..,M) \qquad [4.1]$$

where $\mu(t)$ is the temporal trend or the climate change signal and $E(\mathbf{s})$ is the spatial trend. We emphasize that this spatio-temporal model for expected values is different from the traditional models used in geostatistics or the multivariate statistical methods. As regards geostatistics, purely spatial data is assumed in general.

4.2.3. *Linear regression formula*

In essence, the multiple linear regression formula is the theoretical basis of the different linear interpolation methods. The linear regression between predictand $Z(\mathbf{s}_0,t)$ and predictors $\mathbf{Z}(t)$ can be written as:

$$\hat{Z}_{LR}(\mathbf{s}_0,t) = \mathrm{E}(Z(\mathbf{s}_0,t)) + \mathbf{c}^{\mathrm{T}}\mathbf{C}^{-1}(\mathbf{Z}(t) - \mathrm{E}(\mathbf{Z}(t))) \qquad [4.2]$$

and $\hat{Z}_{LR}(\mathbf{s}_0,t)$ is the best linear estimation that minimizes the mean-squared prediction error. Consequently, the linear regression formula would be the optimal linear interpolation formula concerning the mean-squared prediction error. However, with respect to the application, problems arise from the unknown statistical parameters $\mathrm{E}(Z(\mathbf{s}_0,t))(i = 0,...,M)$ and $\mathbf{c}$, $\mathbf{C}$. Let us assume that the meteorological model [4.1] for the expected values of formula [4.2] can be written as:

$$\hat{Z}_{LR}(\mathbf{s}_0,t) = (\mu(t) + E(\mathbf{s}_0)) + \mathbf{c}^{\mathrm{T}}\mathbf{C}^{-1}(\mathbf{Z}(t) - (\mu(t)\mathbf{1} + \mathbf{E})) \qquad [4.3]$$

where $\mathbf{E}^{\mathrm{T}} = [E(\mathbf{s}_1),.....,E(\mathbf{s}_M)]$ and $\mathbf{1}^{\mathrm{T}} = [1,.....,1]$ are identical. As can be seen, the main problem is the estimation of the unknown climate change signal $\mu(t)$ if we want to apply the optimal linear regression interpolation formula.

4.3. Geostatistical interpolation methods

The various geostatistical interpolation formulae can be obtained from the linear regression formula [4.2] by applying of the generalized least-squares estimation for the expected values. The type of kriging formulae depends on the model assumed for the expected values.

4.3.1. *Ordinary kriging formula*

The ordinary kriging formula is a special case of the universal kriging formula. The assumed model for the expected values is $\mathrm{E}(Z(\mathbf{s}_i,t)) \equiv \mu(t)$ $(i = 0,..., M)$ thus there is no spatial trend. The generalized least-squares estimation for $\mu(t)$ by using only the predictors $\mathbf{Z}(t)$ may be expressed in the form $\hat{\mu}_{gls}(t) = (\mathbf{1}^{\mathrm{T}}\mathbf{C}^{-1}\mathbf{1})^{-1}\mathbf{1}^{\mathrm{T}}\mathbf{C}^{-1}\mathbf{Z}(t)$. By substituting the estimate $\hat{\mu}_{gls}(t)$ into the linear regression formula [4.2] we obtain the ordinary kriging formula as:

$$\hat{Z}_{OK}(\mathbf{s}_0,t) = \hat{\mu}_{gls}(t) + \mathbf{c}^{\mathrm{T}}\mathbf{C}^{-1}(\mathbf{Z}(t) - \hat{\mu}_{gls}(t)\mathbf{1}) = \sum_{i=1}^{M}\lambda_i Z(\mathbf{s}_i,t), \text{ where} \sum_{i=1}^{M}\lambda_i = 1 \cdot$$

The vector of weighting factors $\boldsymbol{\lambda}^{\mathrm{T}} = [\lambda_1,..,\lambda_M]$ can be written in covariance form:

$$\boldsymbol{\lambda}^{\mathrm{T}} = \left(\mathbf{c}^{\mathrm{T}} + \mathbf{1}^{\mathrm{T}}\frac{(1 - \mathbf{1}^{\mathrm{T}}\mathbf{C}^{-1}\mathbf{c})}{\mathbf{1}^{\mathrm{T}}\mathbf{C}^{-1}\mathbf{1}}\right)\mathbf{C}^{-1}, \qquad [4.4]$$

or equivalently in variogram form:

$$\boldsymbol{\lambda}^{\mathrm{T}} = \left(\boldsymbol{\gamma}^{\mathrm{T}} + \mathbf{1}^{\mathrm{T}} \frac{\left(1-\mathbf{1}^{\mathrm{T}}\boldsymbol{\Gamma}^{-1}\boldsymbol{\gamma}\right)}{\mathbf{1}^{\mathrm{T}}\boldsymbol{\Gamma}^{-1}\mathbf{1}}\right)\boldsymbol{\Gamma}^{-1}. \qquad [4.5]$$

The unknown variogram values $\boldsymbol{\gamma}$, $\boldsymbol{\Gamma}$ which are preferred in geostatistics are modeled according to section 4.3.3.

4.3.2. *Universal kriging formula*

The universal kriging formula is the generalized case of the ordinary kriging formula. The model assumption is that the expected values may be expressed as:

$$\mathrm{E}(Z(\mathbf{s}_i,t)) = \sum_{k=1}^{K} \beta_k(t) x_k(\mathbf{s}_i) \quad (i=0,\ldots,M),$$

which in vector form is:

$$\mathrm{E}(Z(\mathbf{s}_0,t)) = \mathbf{x}^{\mathrm{T}}\boldsymbol{\beta}(t) \;\;, \;\mathrm{E}(\mathbf{Z}(t)) = \mathbf{X}\boldsymbol{\beta}(t),$$

where $\mathbf{x}$, $\mathbf{X}$ are given supplementary deterministic model variables.

The generalized least-squares estimation for the coefficient vector $\boldsymbol{\beta}(t)$ by using only the predictors $\mathbf{Z}(t)$ can be written in the form $\hat{\boldsymbol{\beta}}_{gls}(t) = \left(\mathbf{X}^{\mathrm{T}}\mathbf{C}^{-1}\mathbf{X}\right)^{-1}\mathbf{X}^{\mathrm{T}}\mathbf{C}^{-1}\mathbf{Z}(t)$.

It should be noted that in this way the spatial trend $E(\mathbf{s})$ according to formula [4.1] is also modeled by using only the predictors $\mathbf{Z}(t)$. By substituting the estimates $\mathbf{x}^{\mathrm{T}}\hat{\boldsymbol{\beta}}_{gls}(t)$, $\mathbf{X}\hat{\boldsymbol{\beta}}_{gls}(t)$ into the linear regression formula [4.2] we obtain the universal kriging formula as:

$$\hat{Z}_{UK}(\mathbf{s}_0,t) = \mathbf{x}^{\mathrm{T}}\hat{\boldsymbol{\beta}}_{gls}(t) + \mathbf{c}^{\mathrm{T}}\mathbf{C}^{-1}\left(\mathbf{Z}(t) - \mathbf{X}\hat{\boldsymbol{\beta}}_{gls}(t)\right) = \sum_{i=1}^{M} \lambda_i Z(\mathbf{s}_i,t), \text{ where } \boldsymbol{\lambda}^{\mathrm{T}}\mathbf{X} = \mathbf{x}^{\mathrm{T}}.$$

The vector of weighting factors $\boldsymbol{\lambda}^{\mathrm{T}} = [\lambda_1,..,\lambda_M]$ can be written in covariance form:

$$\boldsymbol{\lambda}^{\mathrm{T}} = \left\{\mathbf{c} + \mathbf{X}\left(\mathbf{X}^{\mathrm{T}}\mathbf{C}^{-1}\mathbf{X}\right)^{-1}\cdot\left(\mathbf{x} - \mathbf{X}^{\mathrm{T}}\mathbf{C}^{-1}\mathbf{c}\right)\right\}^{\mathrm{T}}\mathbf{C}^{-1},$$

or equivalently in variogram form:

$$\boldsymbol{\lambda}^{\mathrm{T}} = \left\{\boldsymbol{\gamma} + \mathbf{X}\left(\mathbf{X}^{\mathrm{T}}\boldsymbol{\Gamma}^{-1}\mathbf{X}\right)^{-1}\left(\mathbf{x} - \mathbf{X}^{\mathrm{T}}\boldsymbol{\Gamma}^{-1}\boldsymbol{\gamma}\right)\right\}^{\mathrm{T}}\boldsymbol{\Gamma}^{-1}.$$

The unknown variogram values $\boldsymbol{\gamma}$, $\boldsymbol{\Gamma}$ which are preferred in geostatistics are modeled according to section 4.3.3.

4.3.3. *Modeling of unknown statistical parameters in geostatistics*

In geostatistics only the predictors $Z(\mathbf{s}_i,t)$ ($i = 1,...,M$) constitute the usable information or the sample for the modeling of variogram values $\boldsymbol{\gamma}$, $\boldsymbol{\Gamma}$. It means that we only have a single realization in time for the modeling of the statistical parameters in question. In order to solve the problem of absence of temporal data some assumptions about the statistical structure are made which constitute a simplification of the problem. For example, such assumptions are the intrinsic stationarity or second-order (weak) stationarity, semivariogram $\gamma(Z(\mathbf{s}_i), Z(\mathbf{s}_j)) = \gamma(\mathbf{s}_i - \mathbf{s}_j)$, etc.

4.4. Meteorological interpolation

Similarly to the geostatistical interpolation formulae, an appropriate meteorological interpolation formula can be obtained from the linear regression formula [4.2] by applying the generalized least-squares estimation for the expected values. The key question is the model assumption for the expected values.

4.4.1. *Meteorological interpolation formula*

The meteorological model [4.1] is assumed to be $\mathrm{E}(Z(\mathbf{s}_i,t)) = \mu(t) + E(\mathbf{s}_i)$ ($i = 0,..,M$), where $\mu(t)$ is the temporal trend and $E(\mathbf{s})$ is the spatial trend. By supposing that the spatial trend $E(\mathbf{s})$ is known we apply the generalized least-squares estimation for the temporal trend $\mu(t)$ by using the predictors $\mathbf{Z}(t)$ and the spatial trend $\mathbf{E}^{\mathrm{T}} = [E(\mathbf{s}_1),\ldots, E(\mathbf{s}_M)]$. In this case the generalized least-squares estimate can be written as: $\hat{\mu}_{gls}^{E}(t) = \left(\mathbf{1}^{\mathrm{T}}\mathbf{C}^{-1}\mathbf{1}\right)^{-1}\mathbf{1}^{\mathrm{T}}\mathbf{C}^{-1}(\mathbf{Z}(t)-\mathbf{E})$. By substituting the estimate $\hat{\mu}_{gls}^{E}(t)$ into the linear regression formula [4.3] which is rewritten from formula [4.2] according to the model [4.1] we obtain the following interpolation formula:

$$\hat{Z}_{MI}(\mathbf{s}_0,t) = \left(\hat{\mu}_{gls}^{E}(t) + E(\mathbf{s}_0)\right) + \mathbf{c}^{\mathrm{T}}\mathbf{C}^{-1}\left(\mathbf{Z}(t) - \left(\hat{\mu}_{gls}^{E}(t)\mathbf{1} + \mathbf{E}\right)\right) =$$
$$= E(\mathbf{s}_0) + \sum_{i=1}^{M} \lambda_i \left(Z(\mathbf{s}_i,t) - E(\mathbf{s}_i)\right), \quad \text{where} \quad \sum_{i=1}^{M} \lambda_i = 1 \cdot \qquad [4.6]$$

The vector of weighting factors $\boldsymbol{\lambda}^{\mathrm{T}} = [\lambda_1,\ldots,\lambda_M]$ can be written equivalently in covariance and variogram form according to [4.4] and [4.5]. The obtained interpolation formula is a detrended or residual interpolation formula that includes the spatial trend and the theoretical ordinary kriging weighting factors. However, it is not identical to the detrended or residual interpolation method because the interpolation formula, with the modeling methodology of the necessary statistical parameters together defines an interpolation method. For example, at the detrended interpolation methods applied in practice the modeling of the statistical parameters is based on only the predictors $Z(\mathbf{s}_i,t)$ ($i = 1,\ldots, M$).

4.4.2. *Possibility of modeling unknown statistical parameters in meteorology*

According to [4.6] where the sum of weighting factors is equal to one we have the following appropriate meteorological interpolation formula:

$$\hat{Z}_{MI}(\mathbf{s}_0,t) = \sum_{i=1}^{M} \lambda_i \left(E(\mathbf{s}_0) - E(\mathbf{s}_i)\right) + \sum_{i=1}^{M} \lambda_i Z(\mathbf{s}_i,t),$$

where $\sum_{i=1}^{M} \lambda_i = 1$ and the covariance form of the weighting factors is defined by [4.4].

Consequently, the unknown statistical parameters are as follows: spatial trend differences $E(\mathbf{s}_0) - E(\mathbf{s}_i)$ ($i = 1,\ldots, M$) and covariances $\mathbf{c}$, $\mathbf{C}$. In essence, these parameters are climate parameters which means that we could interpolate optimally if we knew the climate. The special possibility in meteorology is to use the long meteorological data series for the modeling of the climate statistical parameters in question. The data series make it possible to know the climate in accordance with the fundamentals of climatology!

4.4.3. *Difference between geostatistics and meteorology*

The main difference can be found in the amount of information used for modeling the statistical parameters. In geostatistics the usable information or the sample for modeling is only the predictors $Z(\mathbf{s}_i,t)$ $(i=1,...,M)$ which refer to a fixed instant of time, that is, a single realization in time. "Statistically speaking, some further assumptions about Z have to be made. Otherwise, the data represent an *incomplete* sampling of a *single* realization, making inference impossible" ([CRE 91], p. 53). The assumptions are, for example, intrinsic stationarity or second-order (weak) stationarity, semivariogram $\gamma(Z(\mathbf{s}_i),Z(\mathbf{s}_j))=\gamma(\mathbf{s}_i-\mathbf{s}_j)$, which need some simplification in order to solve the problem of absence of temporal data, whereas in meteorology we have space-time data, namely the long data series which form a sample both in time and space and make it possible to model the climate statistical parameters in question. If the meteorological stations $\mathbf{S}_k$ $(k=1,..,K)$ $(\mathbf{S}\in D)$ have long data series, then spatial trend differences $E(\mathbf{S}_k)-E(\mathbf{S}_l)$ $(k,l=1,...,K)$ as well as the covariances $\text{cov}(Z(\mathbf{S}_k),Z(\mathbf{S}_l))$ $(k,l=1,...,K)$ can be estimated statistically. Consequently, these parameters are essentially known and provide much more information for modeling than the predictors $Z(\mathbf{s}_i,t)$ $(i=1,...,M)$ only.

4.5. Software and connection of topics

Our MISH (Meteorological Interpolation based on Surface Homogenized Data Basis) method for the spatial interpolation of surface meteorological elements was developed [SZE 04, SZE 05] according to the mathematical background that is outlined in section 4.4. This is a meteorological system not only with respect to the aim but with respect to the tools as well. It means that using all the valuable meteorological information – e.g. climate and possible background information – is required.

The software MISHv1.01 consists of two units that are the modeling and the interpolation systems. The interpolation system can be operated on the results of the modeling system.

In the following text we briefly summarize the most important facts about these two units of the developed software:

– Modeling system for climate statistical (deterministic and stochastic) parameters:

- based on long homogenized data series and supplementary deterministic model variables. The model variables may be height, topography, distance from the sea etc. Neighborhood modeling, correlation model for each grid point,

- benchmark study, cross-validation test for interpolation error or representativity,

- modeling procedure must be executed only once before the interpolation applications!

– Interpolation system:

- additive (e.g. temperature) or multiplicative (e.g. precipitation) model and interpolation formula can be used depending on the climate elements,

- daily, monthly values and many years' means can be interpolated,

- few predictors are also sufficient for the interpolation and it is not a problem if the greater part of the daily precipitation predictors is equal to 0,

- the interpolation error or representativity is modeled too,

- capability for application of supplementary background information (stochastic variables), e.g. satellite, radar, forecast data.

As can be seen, the modeling of the climate statistical parameters is a key issue to the interpolation of meteorological elements and modeling can be based on the long homogenized data series. The necessary homogenized data series can be obtained by our homogenization software MASHv3.01 (Multiple Analysis of Series for Homogenization; [SZE 99, SZE 06]). Similarly to the connection of interpolation and homogenization, in our conception the meteorological questions cannot be treated separately. We present a block diagram (Figure 4.1) to illustrate the possible connection between various important meteorological topics.

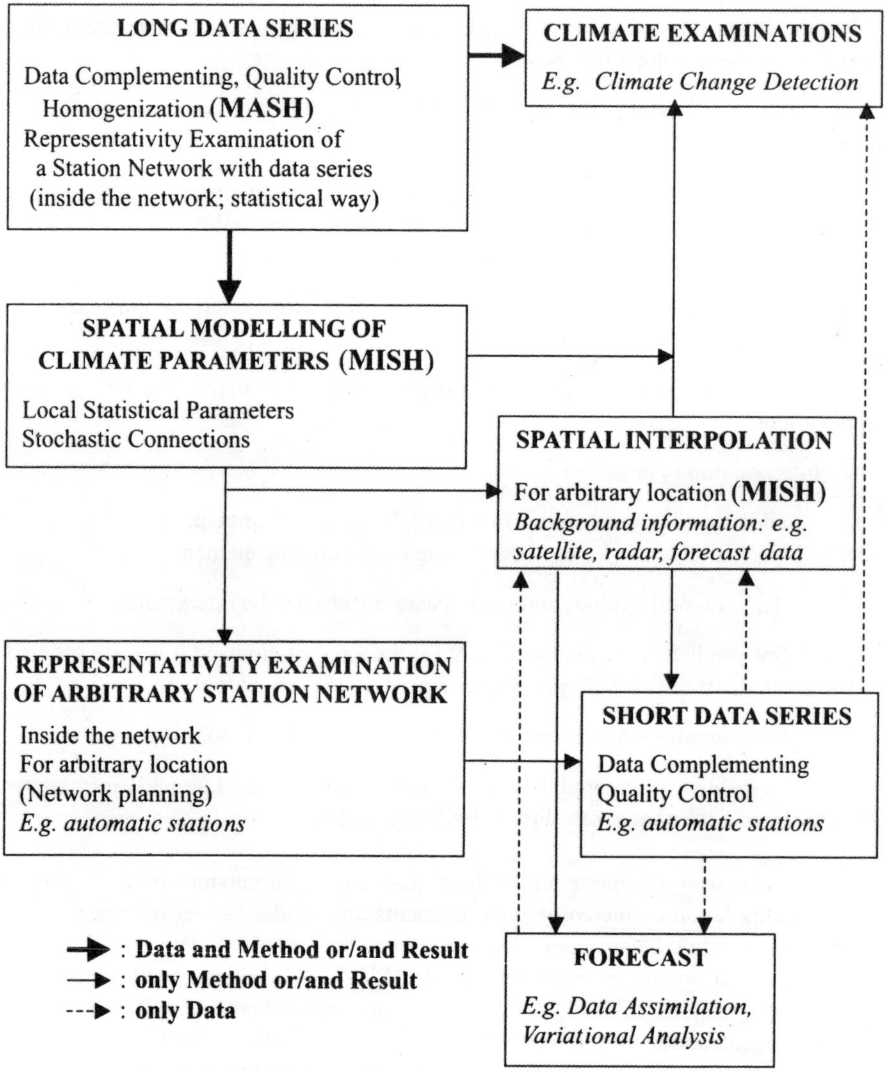

Figure 4.1. *Connection of topics and systems*

4.6. Example of the MISH application

Finally, we present an example of the application of the MISH system. The example is related to the interpolation of the daily precipitation sum in extreme meteorological situation. The interpolation is performed with and without the application of supplementary background information. The space domain is Hungary with 0.5'x 0.5' resolution, the predictors are the surface daily precipitation sums from 661 stations and the background information used is the gauge-adjusted 24 hours radar precipitation data. We do not detail the multiplicative interpolation formula used for the precipitation sum or the mathematical methodology developed for the application of background information since it would be beyond the scope of this chapter. Only the interpolation maps (Figure 4.2) are shown.

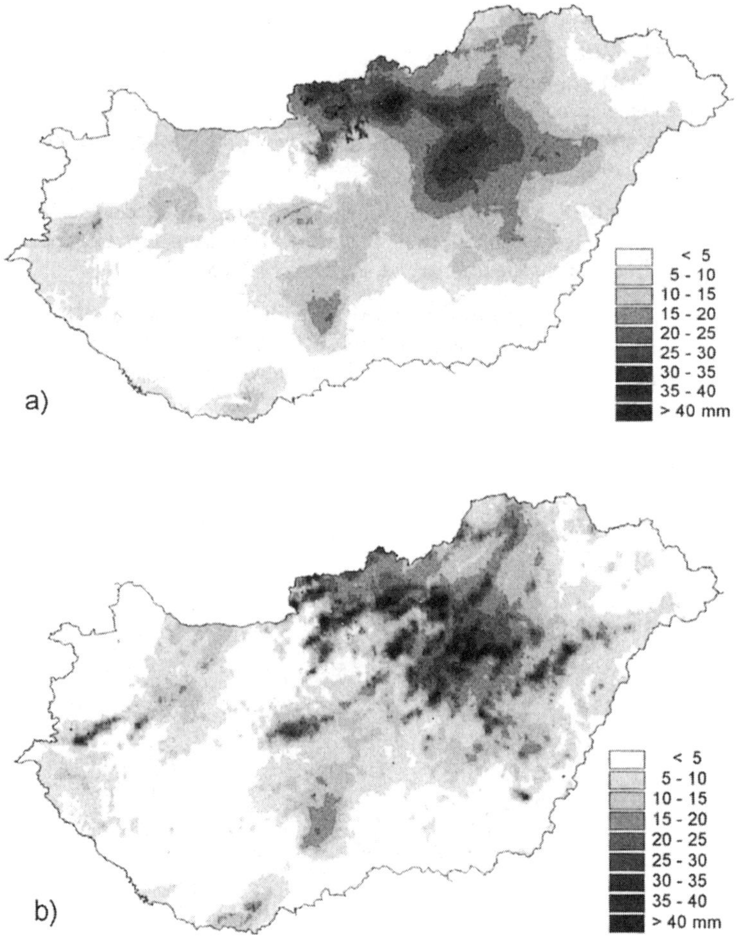

Figure 4.2. *Interpolation of daily precipitation sum without (a) and with (b) radar data as background information (Hungary, 29.07.2006)*

4.7. Bibliography

[CRE 91] CRESSIE N., 1991: "Statistics for Spatial Data", Wiley, New York.

[SZE 99] SZENTIMREY T., 1999: "Multiple Analysis of Series for Homogenization (MASH)", Proceedings of the Second Seminar for Homogenization of Surface Climatological Data, Budapest, Hungary; WMO, WCDMP-No. 41, pp. 27-46.

[SZE 02] SZENTIMREY T., 2002: "Statistical problems connected with the spatial interpolation of climatic time series", http://www.knmi.nl/samenw/cost719/documents/Szentimrey.pdf.

[SZE 04] SZENTIMREY T., BIHARI Z., 2004: "Mathematical background of the spatial interpolation methods and the software MISH (Meteorological Interpolation based on Surface Homogenized Data Basis)", Proceedings of the Conference on Spatial Interpolation in Climatology and Meteorology, COST-719 Meeting, Budapest, Hungary, 24-29 October 2004.

[SZE 05] SZENTIMREY T., BIHARI Z., SZALAI S., 2005: "Meteorological Interpolation based on Surface Homogenized Data Basis (MISH)", European Geosciences Union, General Assembly 2005, Vienna, Austria, 24-29 April 2005.

[SZE 05] SZENTIMREY T., BIHARI Z., SZALAI S., 2005: "Limitations of the present GIS methods in respect of meteorological purposes", 5th Annual Meeting of the European Meteorological Society (EMS)/7th ECAM, Utrecht, the Netherlands, 12-16 September 2005.

[SZE 05] SZENTIMREY T., BIHARI Z., 2005: "Manual of interpolation software MISHv1.01".

[SZE 06] SZENTIMREY T., 2006: "Manual of homogenization software MASHv3.01".

Chapter 5

Uncertainty from Spatial Sampling: A Case Study in the French Alps

5.1. Introduction

The goal of any environmental research is often to gain more precision in its results, thus paying attention to the different sources of imprecision coming from:

– the differential validity of the observations;

– their combination in a model;

– their temporal distribution;

– their spatial location.

We emphasize this last point, since a climatic map interpolates point values, which implies a representative sample. The question then is:

– *"is it representative of the whole territory?"*

– *"is it equally representative of the whole territory?"*

– *"is it representative of something else, for instance, population distribution?"*

Our case study deals with the spatial sample of Lhotellier's PhD work (Figure 5.1). We only consider here the position in latitude, longitude and altitude of 168 weather stations in the northern part of the French Alps (≈ 25,000 km^2) without any reference to measured values.

Chapter written by Pierre DUMOLARD.

There are quite a lot of methods to deal with the subject of spatial samples (Figure 5.2), most of them coming from parametric statistics (thus implying strong assumptions). In the context of COST 719, we decided to apply a GIS exploratory approach (Figure 5.3) to the spatially differential validity of a spatial sample (namely the 168 weather stations in the northern French Alps).

The first attempt considers the sample as a whole, the second one is more local, dealing with a 150m pixel resolution and the third tries to summarize, at two scales, the regional representativeness of the sample.

5.2. The sample as a whole

Firstly, we used some simple exploratory techniques to verify if the sample is representative.

Looking at a map: 2D perception

Figure 5.1 shows the spatial distribution of the weather stations (overlaid on a DEM).

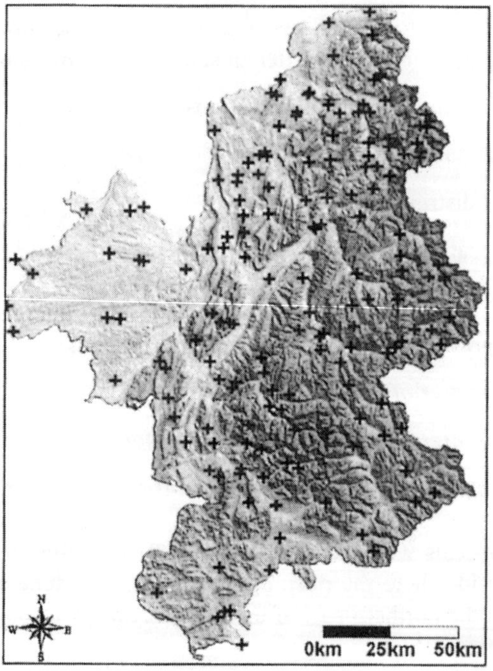

Figure 5.1. *Distribution of 168 weather stations (source: R. Lhotellier & Meteo France)*

A first glance at this map suggests that the measurement points are not evenly distributed in latitude (Y) and longitude (X). To verify XY independence, we can calculate a regression.

A simple linear regression between X and Y for the 168 points gives a R^2 of 0.05. The probability α of a false rejection of independence between X and Y is 0.044. This means that given a weak ($\alpha = 0.01$) risk, we reject independence between latitude and longitude but we accept it when given a larger but still reasonable ($\alpha = 0.05$) risk. The conclusion is doubtful: some further attempt has to be made in order to decide matters.

The calculation of *Ripley's R* (Figure 5.4) and its significance test confirm that the sample is not fully representative of the *whole* area. Ripley's R is based on the calculation of the 2D nearest neighbor of each of the 168 points.

Ripley's R values:

– < 1 indicates a trend towards a concentrated sample;

– ~ 0 indicates hazard distribution;

– > 1 indicates a trend towards a regular (systematic) sample.

The calculation of Ripley's R tells us, here, that the sample as a whole is significantly *concentrated.* By looking at Figure 5.1, we hypothesize that the measure points are concentrated in the mountainous areas and rarer elsewhere (where spatial interpolation is less uncertain). To test R, let us call:

– d_1 the average Euclidean distance of the measure points;

– d_0 the theoretical mean distance (given by Poisson's law);

– σ_{d0} its theoretical standard deviation.

Since N is > 30, we can apply the ε test based on the Laplace-Gauss probability law where

$$\varepsilon = | d_1 - d_0|/\sigma_{d0} = 5.35$$

Since ε is > 2.58 ($\alpha = 0.01$) we reject the hypothesis that the 168 weather stations are a random sample of the area. However, is it not a "*map optical illusion*" due to the *2D* areas representation?

Introducing altitudes

Figure 5.2 investigates the representativeness of the altitude of the sample: the cumulative curves of altitudes show quite a big cumulative difference between the altitudes of the DEM and those of the sample, mainly over 2,000 meters high (exactly where temperature interpolation is more difficult). On the right, Figure 5.2 locates their main difference areas.

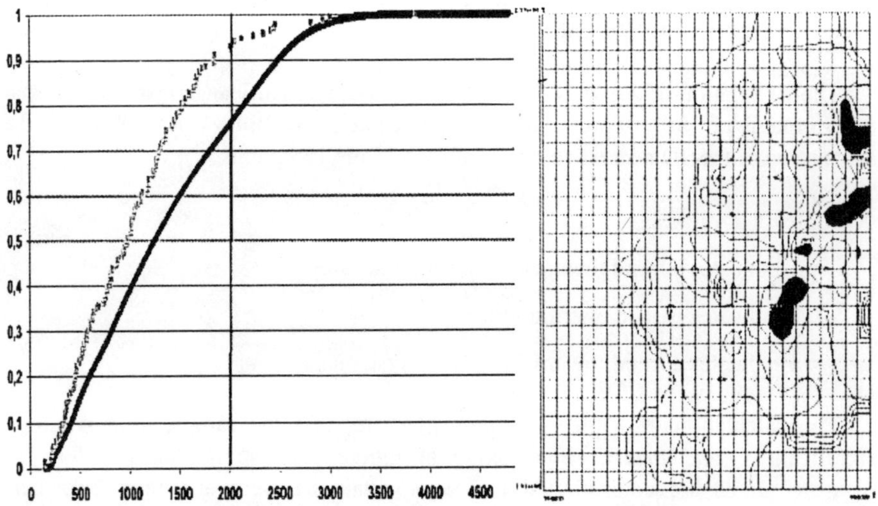

Figure 5.2. *Altitude representativeness of the sample (source: R. Lhotellier & Meteo France)*

In order to verify the visual feeling of 3D non-independence, we have calculated a multiple linear regression:

Altitude = f (longitude, latitude).

$R^2 = 0.41$ (with a very low α risk) means that we have to reject the hypothesis of their independence in *3D* space.

In fact, the sample is not fully representative of the *physical area*, mainly because most of the 168 measure points are situated under 2,000 meters but this is exactly where population and activities are: the sample may be representative of their *vulnerability* to climate.

5.3. Looking in detail where the sample is not representative

In order to consider the sample in detail (at a local scale of 2.25 ha per pixel), we have:

– simplified the original DEM (from a resolution of 50 m to a resolution of 150 m);

– calculated for each 150m pixel its main slope (in decimal degrees);

– transformed decimal degrees in radians (radian = degree * $\pi/180$);

– calculated the 3D increase of the area, i.e. s3d = 1/cosine(slope) for each pixel.

A first map could represent the absolute difference between the 3D ("real") and 2D (plane projection) area of each pixel. We have chosen to show (Figure 5.3) their relative difference.

For each 100 m elevation class (there are 47 classes from 100 to 4,800 m), we have calculated:

– DIF_{dem}: 3D – 2D area for the DEM;

– DIF_{sample}: 3D – 2D area for the sample;

– $(DIF_{dem} - DIF_{sample})/DIF_{dem}$;

– expressed it on a scale [–100% to +100%], 0% being the average of the map.

Thus, Figure 5.3 expresses the relative over- or under-representativeness of the sample in 3D space.

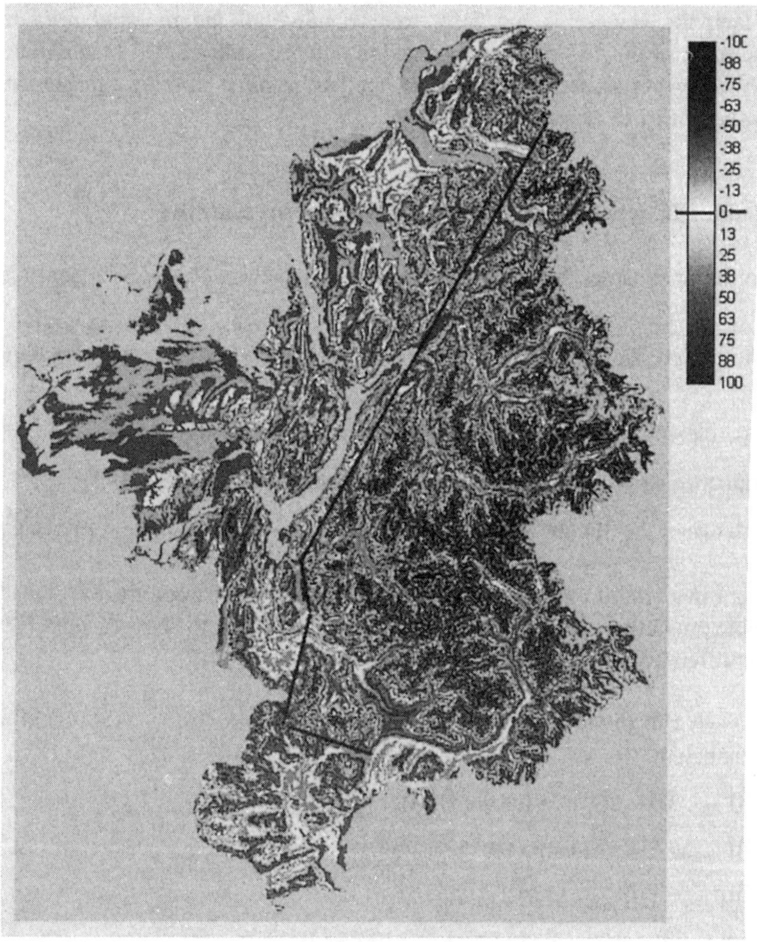

Figure 5.3. *Relative 3D representativeness of the sample at local scale*

Globally, the sample is not a representative one in 3D space and it is still less representative in the most alpine (east) part of the region. There, in detail, the massifs are very much underrepresented but there is not the case for the main valleys where most people live and most economic activities take place.

Among the massifs, the Ecrins (south mountainous area) is the most compact one (weaker glacier erosion in the quaternary period) and so has a bad representation, whereas the Mont Blanc massif (northern part of the mountainous area) is less underrepresented (stronger glacier erosion in the past and therefore larger valleys). The Vanoise massif (main ski resorts in the French Alps) is situated between them.

In the mountainous massifs, the climatic interpolation is more difficult but they have few inhabitants and scarce economic activity, except tourism (which is the only way to maintain inhabitants and spatial infrastructures there).

5.4. Summarizing the sampling uncertainty

Even in a GIS exploratory context only, there are several ways to summarize the 2D and 3D differential validity of a spatial point sample.

5.4.1. *2D simplification*

One simple idea is to create buffer zones around each measurement point: the areas inside the buffers may be considered as more reliable than those outside. However, the size of the buffers is unknown: it is possible to increase it step by step, by producing something like a cartographic "animated cartoon". Figure 5.4 is an illustration for only 3 buffer sizes: 1.5, 3 and 5 km which underline each time the main areas of uncertainty.

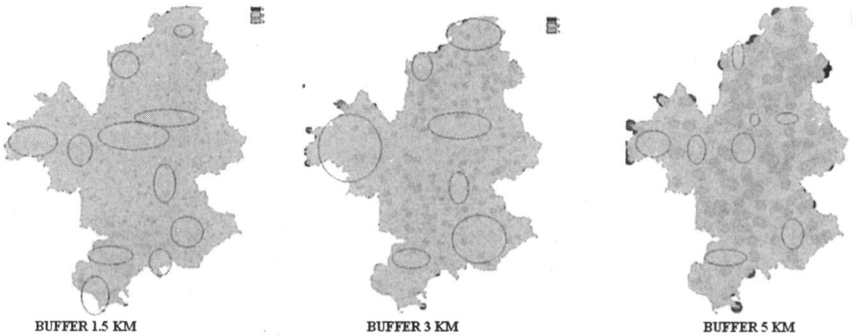

Figure 5.4. *Main 2D areas of uncertainty when increasing the size of buffers*

This view is a bit too binary and over-simplistic but could be seen as a first step towards spatial generalization of the 2D spatial uncertainty.

More complex approaches exist and, among them, spatial scan techniques (Figure 5.5) can be very useful, for instance, in the search for *spatial hierarchical clusters*. The so-called "hot spots" algorithm is a probabilistic extension of the nearest neighbor method using a hierarchical classification tree (such as a Ward classification). The data file contains latitude and longitude values only: sample points are gradually grouped according to their nearest neighbor distance. At each step of the procedure, the intragroup distance variance is calculated for each cluster: if the variance becomes too big its creation is stopped so that, in the end, not all points are classified. The convex hulls surrounding classified points ("*hot spots*") are considered better sampled areas than those outside and a certainty index for the hot spots may be built depending on their variances. A global index of certainty/uncertainty can be derived from a simple variance analysis (if the main constraints of the model are reasonably respected). Other statistical tests can be carried out.

We have tried the "hot spot" algorithm using the *Crimestat* freeware but it gave results which were strongly dependent on edge effects either because we were unfamiliar with the software or because it could not take them properly into account.

5.4.2. *3D generalization*

A first generalization can be done using the Voronoi (or Thiessen) tessellation. We have calculated the 168 Voronoi polygons corresponding to our 168 measure points. Each polygon delimits, in 2D Euclidean space, an area which is nearer to its inner point than any other (since the Delaunay triangulation is the geometric dual of the Voronoi tessellation (see Figure 5.5)).

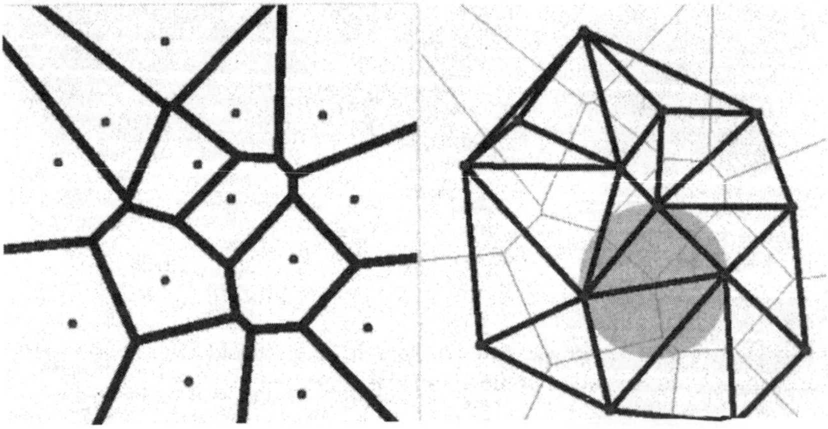

Figure 5.5. *Voronoi tessellation and corresponding Delaunay triangulation*

The calculation of the 168 3D surfaces gives an estimate of the accuracy of parameter interpolation inside each polygon: a large 3D surface means a weak accuracy, a small surface means good accuracy and contiguous small surfaces delineate the reliable regions. However, we must consider not only the average (m_j) 3D surface of each polygon but also the standard deviation (σ_j) of the 3D surface of its pixels. Figure 5.6 is the map corresponding to σ_j/m_j, ($j = 1,\ldots, 168$): a low value means XYZ homogenity of the polygon, a larger value means heterogenity and thus a weaker reliability for parameter interpolation.

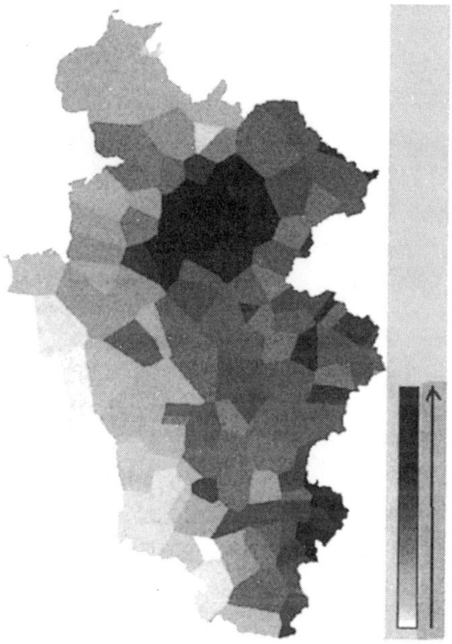

Figure 5.6. *XYZ homogenity of the Voronoi polygons (indicating their reliability)*

Obviously, because of strong differences in height, the alpine part is more heterogenous: there clearly is a global west-east gradient of reliability. In more detail, the main massifs (Ecrins, Mont Blanc with summits of over 4,000 m) appear as the least reliable areas.

Another way of doing this is to summarize, inside the Voronoi polygons, the pixel values of Figure 5.3 (scaled [–100 to 100]), where negative values mean relative poor representativeness and positive values indicate an above average representativeness. As polygons are aggregates, we now deal with distributions of 3D areas and not single values. We therefore need a tool to simplify and compare the 168 distributions.

To simplify them, we have divided the [–100 to 100] continuums in 8 equal intervals classes. Counting, for each polygon, the number of pixels in each class creates a contingency table the size of which has been reduced by a correspondence factor analysis. The 3 first factors accounted for 95% of the initial information. Moreover, they are independent (additive) super variables: it enabled us to create clusters, without any correlation bias.

The cluster analysis of 168 polygons and 3 factors gave good results for 7 groups (intergroup variance = 90.5% of total variance, intragroup variance < 10%). The 7 groups of Voronoi polygons are globally homogenous and different from each other. Figure 5.7 represents their mapping and profiles (reduced to 4 "quarters").

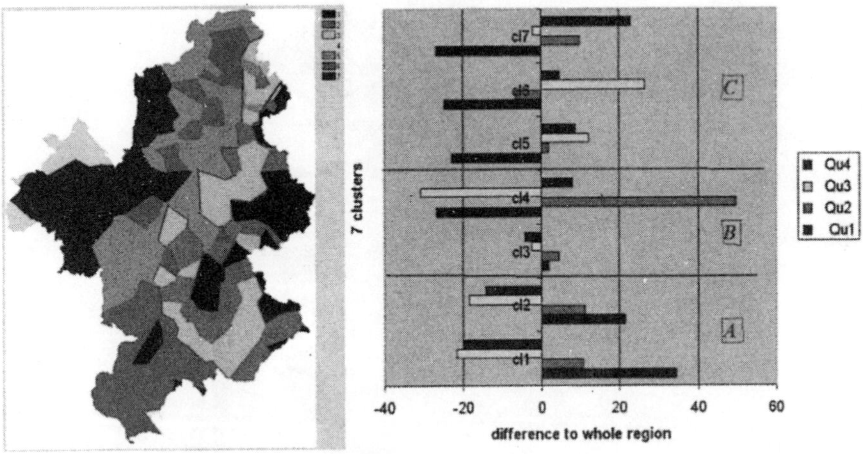

Figure 5.7. *Map and profiles of the 7 groups of Voronoi polygons*

There is a clear gradient from WNW to ESE, meaning that (considering 3D surfaces) the sample over-represents plains and hills (north west crescent) and under-represents mountain areas (east crescent). This observation, which had already been made, confirms the benefit to take into account 3D surfaces rather than projected 2D surfaces.

5.4.3. *Geographic homogenous sub-regions of the sample*

If we build a table of 1,085,214 rows (pixels) and 3 columns (latitude, longitude, altitude), principal components and cluster analysis results in 10 spatially homogenous sub-regions. For each of these 10 we have calculated two criteria of good or bad sub-samples found inside weather stations:

– their sampling *density* (number of stations per 1,000 km²): a high density indicates a better sub-sample, whereas a low density indicates a less representative one;

– the *arrangement* of points inside each sub-region, using Ripley's R and its t student test (since the number of points by sub-region is less than 30). Let us recall that a R value < 1 represents tendency to points agglomeration, a value ≈ 1 represents a hazard distribution and a value > 1 trend to systematic pattern. Some sub-regions have fairly good sub-samples. If they were a majority, the stratification in geographic sub-regions would have been successful.

By combining these two indices, we get Table 5.1.

	Concentrated pattern	Random pattern	Regular pattern
Low density	no. 3, no. 7	no. 4	no. 5
Average density	no. 9	no. 1, no. 10	no. 6
High density	–	–	no. 2, no. 8

Table 5.1. *Combination of density of points and type of pattern for the 10 sub-regions*

Figure 5.8 is a map corresponding to Table 5.1.

Meaning of its hierarchical legend:

– *category 1*: sub-region with a low density of points *and* a concentrated pattern;

– *category 2*: sub-region with a low density of points;

– *category 3*: sub-region with a concentrated pattern;

– *category 4*: sub-region with correct density and pattern of points;

– *category 5*: sub-region with a good sample (density + pattern).

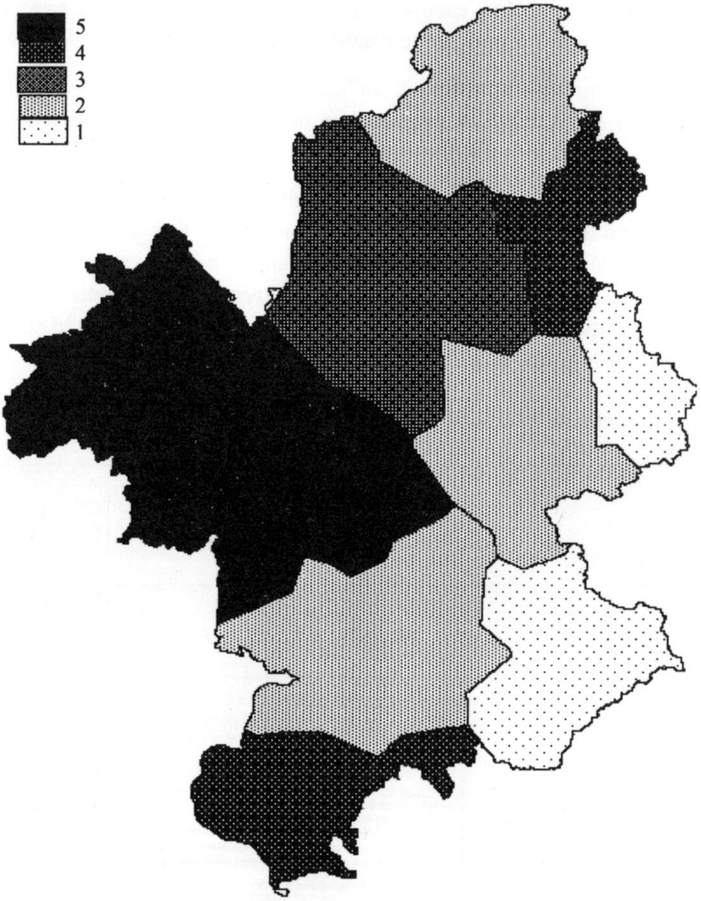

Figure 5.8. *Simplified quality of the sample by sub-region*

5.4.4. *Interpolation of a climate parameter*

This cluster approach of a spatial point sample defines a spatially stratified sample where strata are homogenous according to latitude, longitude and altitude. The quality of the sample in each stratum is (quantitatively or qualitatively) known and it is then possible to interpolate a climate parameter (temperature, for instance) *in each one*. Since each stratum is more homogenous than the global sample, the assumption of an *isotropic space* (implied by most spatial analysis interpolation techniques) is better satisfied.

This approach delivers 3 pieces of information:

– the local interpolated values of the parameter (the corresponding map may be smoothed if the parameter is spatially continuous);

– the representativeness of the sample in each stratum (which can be smoothed and mapped superimposed to the interpolated values);

– the global quality of the interpolation through an analysis of variance (which can establish if the stratification improves the interpolation or not).

5.5. Conclusion

Whether a point sample is generally representative of a whole area is a matter to examine previously to any interpolation: it generally gives confidence in its regionalized results. Especially in heterogenous regions (like the Alps) at any scale, the use of traditional "spatial analysis" techniques oversimplifies the research of differences in the sample quality since these techniques are based on the assumption of an isotropic 2D Euclidean space (which is clearly not fulfilled in this case).

The use of GIS and exploratory statistics may be a better approach, which makes it possible to consider 3D surfaces, to easily test several solutions (according to the scale of the test) and to summarize their results in a (post-)stratification of the sample. The point density of the strata as well as their spatial inner arrangement may be seen as quantitative estimations of their reliability, the combination of these two indices giving a qualitative one.

The map of this qualitative index may be superimposed to an interpolated parameter (for example, temperature) map, thus showing both probable value and reliability. Where sample reliability is too small, no value might be displayed.

If the point sample changes over time, we then face a problem of higher complexity but in any case the leading idea is the same: interpolation of a parameter is not feasible everywhere at any time.

It can be said that interpolating values from a point sample is not feasible *always and everywhere.* This makes a big difference to the interpolation techniques existing in most GIS.

5.6. Bibliography

[BER 68] BERRY B.J.L., BAKER A.M., "Geographic sampling" in Berry B.J.L. & Marble D.F., "Spatial analysis: a reader in statistical geography", Prentice Hall, 1968.

[CRE 93] CRESSIE N., "Statistics for spatial data", Wiley & Sons, 1993.

[FOT 00] FOTHERINGHAM A.S. *et al.*, "Quantitative Geography", Sage, 2000.

[GET 96] GETIS A., ORD J.K., "Local spatial statistics: an overview" in Longley P. & Batty M., "Spatial analysis: modelling in a GIS context", Wiley & Sons, 1996.

[LHO 06] LHOTELLIER R., "Spatialisation des temperatures en zone de montagne alpine", PhD Thesis, Joseph Fourier University, 2006-06-21.

[RIP 91] RIPLEY B.D., "Statistical inference for spatial process", Cambridge University Press, 1991.

[ZAN 05] ZANINETTI J.M., "Statistique spatiale, méthodes et applications géomatiques", Hermes, 2005.

Part 2

Spatial Interpolation of Climate Data

Chapter 6

The Developments in Spatialization of Meteorological and Climatological Elements

6.1. Introduction

An important but difficult task for a climatologist or a meteorologist is to provide information about weather and climate for any place at any time, even for places where observations of the meteorological elements do not exist. In such cases they have to use their skill and knowledge to give the most reliable value for the desired information. Traditionally this is done by using observed values at neighboring stations which are then adjusted for representativity, terrain and other effects affecting the local climatology. Such estimates have usually been carried out as single point calculations, often including subjective considerations based on local knowledge and experience. Most of these estimates will not be consistently derived and they are thereby not reproducible and cannot be regarded as homogenous. They are therefore of limited value, for example, for advanced climate analysis.

Geographical Information Systems (GIS) give possibilities to combine different georeferenced variables and parameters in such a way that it should be also possible to give consistently derived estimates of meteorological and climatological variables at any location at any time. This potential was realized by the European Meteorological Services in the mid-1990s, which meant an increasing activity within this field over the last 5-10 years. In the beginning, GIS was primarily used as a tool to establish continuous maps of climate reference values of several

---

Chapter written by Ole Einar TVEITO.

elements. Many of the attempts were carried out by using the built-in, though very limited, possibilities for spatial interpolation.

However, most European NMSs realized that their knowledge in using GIS in the most efficient way, also as a tool in spatial interpolation, was limited and that cooperation was needed. This resulted in a project within the European Climate Support Network (ECSN) [DOB 01], which concluded that the work in developing and testing different spatialization schemes had to be done as well as that of designing algorithms for gridded databases. Beside this, strong connections to (or easier access to) GIS-based information tools with database applications had to be established nationally and internationally. One of the outcomes of the ECSN-activity was to initiate COST Action 719, in order to further assess the issues mentioned above.

In this chapter some of the outcomes of the ECSN GIS and COST719 projects will be presented.

6.2. Spatialization

Firstly, however, the term spatialization should be discussed. What does it actually mean? Spatialization is a term which is not strictly defined in other works. It might stand for: "a set of methods describing the dependency of neighboring data of a dataset in a typically Cartesian co-ordinates system". This is, however, a very general statement, so a more in-depth explanation is needed. Traditionally, the main aims of spatialization have been to condense and visualize data. However, spatialization can be also seen in the wider context of spatial analysis, where it means a transformation to derive new or extended information from existing information. It is in this wider context that the working group 2 of COST 719 has been addressing the topic of spatialization and formed a European expert network on spatialization by using the possibilities offered by GIS to perform the analysis of spatial structures of climatic and meteorological elements for the purpose of spatial inter and extrapolation.

6.3. Why spatialization?

The motivations for using spatialization in climatology and meteorology are several and reflect the manifold use of multi-dimensional climatic and meteorological information for both the user community as for internal climatological/meteorological use. Spatialization plays an important role in the derivation of the spatially continuous fields, adding value to such products by combining and using several georeferenced information sources.

The information demanded by different users spans a variety of elements and scales, thus reflecting the various user applications. This will have implications for the specifications of the applications and the required accuracy of the spatially distributed values.

The largest challenge in spatialization is therefore to balance the available information (both meteorological/climatological data and other geographical information) with the specific needs of the user application. The selection of the proper spatialization scheme for each individual purpose needs thoroughly worked out specifications concerning the requested spatial and temporal scales as well as the desired accuracy.

The motivation and objective of the COST719 working group 2 on spatialization has been to evaluate and work out recommendations for different spatialization applications. The working group was composed by researchers from all over Europe with unique expertise within spatialization. This has made it possible to form an excellent network on spatialization within meteorology and climatology by using GIS and considering theoretical as well as practical issues.

6.4. The role of GIS in developing spatialization within climatology

In the early 1990s the GIS software developed rapidly and included the introduction of a variety of new functionalities which enabled advanced spatial analysis. The release of desk-top GIS like the ESRI ArcView made the access to using GIS easier. At the same time there was a need for all European National Meteorological Services to establish new standard normal values and their corresponding maps. For this purpose many NMSs wanted to apply GIS in order to prepare these maps in an objective and consistent way. At the same time, "Global Climate Change" became an issue, demonstrating the necessity of reliable gridded datasets of basic climate elements for analysis of spatio-temporal variability of climate elements.

The first analyses carried out were quite simple and applied the built-in capacities of the GIS software directly on the data. However, it was obvious for all that this did not produce reliable estimates and that more specialized spatialization approaches were needed for producing spatially distributed climatological information.

Since then a large development of GIS-relevant methods for several climatological elements has been seen. In the following sections some of the major trends in the spatialization of climate elements are presented together with some practical and theoretical aspects that have to be considered in the spatial analyses.

6.5. Methodology

Spatialization methods applied in climatology are all based on basically three principles: deterministic, stochastic and pure mathematical, or a combination of these. The choice of method depends on several factors which are all associated with certain assumptions that have to be fulfilled in order to be applied. However, most spatial interpolation methods can mathematically be reduced to basically the same principles [OBL 82]. Therefore, most of the applications follow the same principles in order to meet the assumptions and criteria required by the methods. These criteria are basically based on the assumption on second order spatial stationarity or, for the geostatistical approach, the intrinsic hypothesis.

The climatic conditions in the different parts of Europe vary dramatically and also the need of information about weather and climate varies considerably. The applications developed throughout Europe will therefore differ from region to region, reflecting the local/regional climate conditions and the information about weather and climate that is relevant for the users. The choice of a spatialization method also depends on the characteristics of the weather element to be described. Some elements, like temperature, describe continuous fields, whereas others, like the occurrence of precipitation, can be described as a discrete binary feature. Such different "nature" of the elements will lead to different challenges developing a robust spatialization method.

In practice, the spatialization of a climatic element has shown to be a two-step procedure:

– "Normalize" the observed *in-situ* data in order to obtain the assumption of spatial stationarity.

– Interpolate the normalized field.

The first point, i.e. to find the best way to normalize the data, is the most challenging. This is often done by applying external information expected to have a significant influence on the climate variable as predictor in a deterministic model. The most applied approach for finding such relations is linear regression models or, when using several external predictors, multiple linear regression. In recent years Artificial Neural Networks (ANN) are also introduced, combining the deterministic and stochastic components.

For the interpolation of the "normalized" (or detrended) field any method can be used. The most preferred methods though are geostatistical methods, inverse distance and various spline techniques.

The precision of spatialization method statistics can be assessed by comparing the estimates with the observed values. There are a few common similarity measures that can be used to validate estimates from almost any spatialization method. It is a common strategy to carry out an independent validation procedure and there are basically two practical approaches for it.

The first and perhaps the best one is to split the data sample into two parts and use these two parts for estimation and validation respectively. This approach is applicable when the original data sample is large and regularly sampled and fulfils the criterion of second order stationarity (intrinsic hypothesis) which is often assumed by the spatialization methods. This criterion is rarely obtained by regular meteorological networks and therefore this method is not often really applicable. It might only be used when having very few independent validation stations, keeping the majority in the dataset used for estimation.

The other approach is cross-validation, which is probably the most widely applied validation method within climatology. This approach considers all the data in the validation process. In a cross-validation procedure, one data point is left out of the data sample at a time. An estimated value for this point is derived by using all the other data points. This procedure is repeated until a value is estimated for all the original data points. One possible drawback of using cross-validation is that the whole data sample is often used to define the interpolation model and therefore the validation might not be considered to be totally independent. However, this consideration is negligible when the dataset used is fairly large.

6.6. Data representativity, quality and reliability

The representativity of the observation network is probably the most serious problem within the spatialization of meteorological and climatological elements. Such networks are usually irregularly distributed in both 2D and 3D. To complicate this issue further, station networks will change over time and this will concern the location of the individual stations as well as the spatial density of measurement sites.

There are several ways to address the irregularity of station networks. The two-dimensional representativity of stations can, for example, be described by calculating Voronoi diagrams (Thiessen polygons) around each station. One example is given in Figure 6.1 where the variation of the surrounding area is shown. This example is taken from the NORDGRID project [JAN 07] and shows quite clearly the different problems that arise with station networks. Here are the merged data networks in four different countries showing, for example, a tremendously dense network in Denmark which is the smallest country and a rather sparse network in Finland. In addition, there is the smallest variance of the terrain in the areas with the densest station network (Figure 6.1).

This means that many of the assumptions underlying the spatialization methods are not completely fulfilled:

– Station networks are usually biased towards lower altitudes and populated areas.

– Station density varies → uncertainty is a function of station density.

– In a sparse network the changes in the network will have consequences for the homogenity of *gridded time series*.

Therefore, the uncertainty will not be a fixed property in spatial analysis, but will vary in space and time.

In meteorology and climatology, terrain characteristics are regarded as one of the main explanatory variables and the input data should represent the same "universe" as the one that shall be estimated. In an ideal world that means that the frequency distribution of the input data and the explanatory variables should coincide. Whether this is true can be investigated by comparing the distribution functions directly, e.g. the altitude distribution of the terrain model and the distribution of station elevation (Figure 6.2). In mountainous areas the stations are usually biased to the detriment of higher elevations.

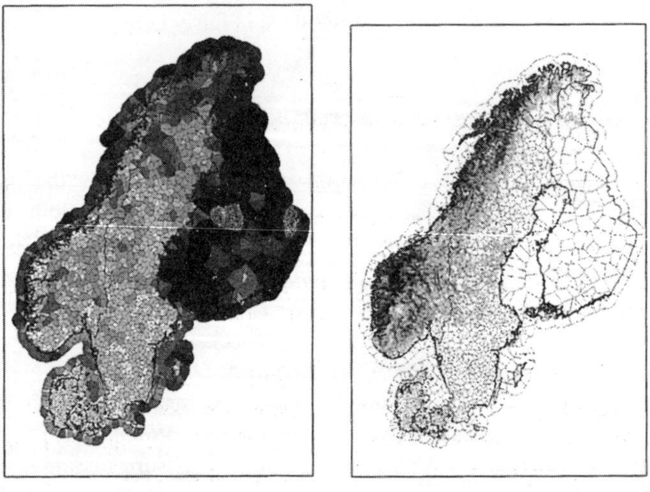

Figure 6.1. *The left panel shows the (area of) Thiessen polygons of the meteorological network used for the NORDGRID project. The right panel shows the standard deviation of the terrain model (dark color for high standard deviation, light color for law standard deviation) indicating the variability of the terrain. Ideally areas with high variability should have higher station density*

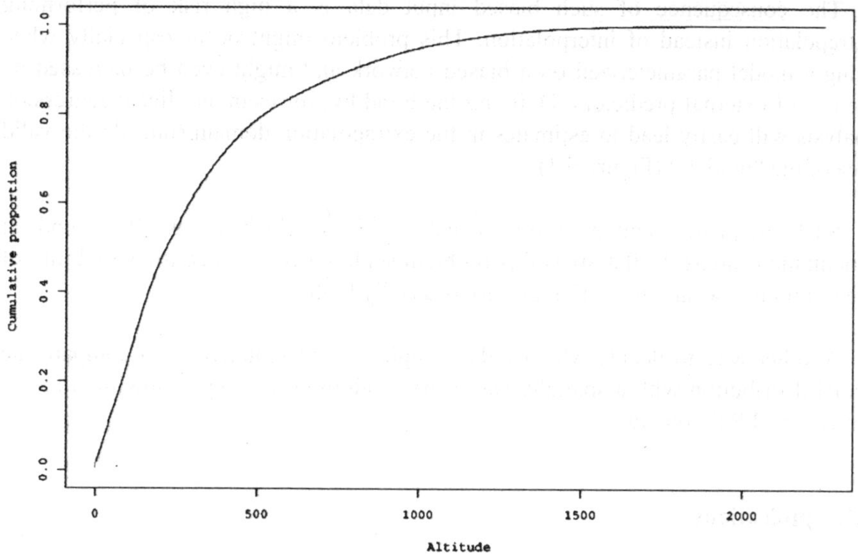

Figure 6.2. *The (cumulative) frequency functions of the altitude of the meteorological network of the Nordic countries shown in Figure 6.1 (gray curve) and the 1 x 1 DEM (GTOPO30, black curve)*

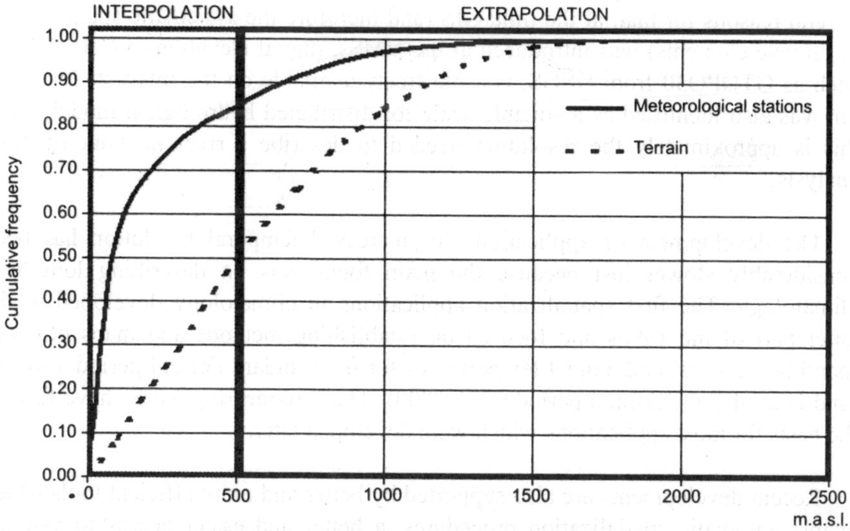

Figure 6.3. *The problem of interpolation vs. extrapolation using a spatially biased dataset with respect to elevation (example from Norway)*

The consequence of such biased input data is a high risk of performing extrapolation instead of interpolation. This problem might occur especially when using a model parameterized on a biased network and might even be increased by the use of external predictors. Defining the trend by, for example, linear regression analysis will easily lead to estimates in the extrapolation domain (outside the valid area of the "model") (Figure 6.3).

Such frequency curves easily reveal the bias which is usually found in mountainous areas. In flat areas this problem seldom occurs since the variability of the terrain is low and the station density is usually high.

Another way to decide whether the sample is representative is to compare the actual distribution with a spatially stationary random process by following a multi-dimensional Poisson law.

6.7. Applications

As mentioned above there has been a huge development in the application of spatialization techniques for deriving spatially distributed climate information during the last decade. What have been interesting to observe over this period are the different developments concerning the spatial and temporal scale. The spatial resolution of the derived dataset was already high at the beginning, typically 1 km x 1 km, whereas the temporal resolution has increased much more slowly. There are several reasons for that: at the time GIS (and therefore the systematic spatialization of climate elements) was introduced at the NMSs, digital elevation models (DEM), such as GTOPO30 from USGS, became freely available on the Internet. 1 km x 1 km was also identified as a suitable scale for distributed hydrological modeling, as this is approximately the resolution needed to describe a river network by flow analysis.

The development of applications for increased temporal resolution has been considerably slower just because the main focus was on describing long-term climatology. The first spatialization applications in climatology developed in the latter part of the 1990s and focused on establishing methods and maps of mean monthly, seasonal and annual temperatures for the standard normal period 1961-90 (and later also the normal period 1971-2000). These pioneering works have formed the basis for most applications which were developed later.

Recent developments are also supported by better and more efficient tools which enable automatic spatialization procedures, a better and easier access to external datasets to be used as explanatory fields and the increased need for such information.

6.7.1. *Spatialization of temperature*

Temperature is a rather easy element to use as an example for spatialization. Contrary to, for example, precipitation it is a continuous element in space and time, and has a strong co-variation with topographical characteristics. It is therefore relatively easy to compensate the bias which is usually present in the observed values to fulfill the assumptions. An approach that has been widely applied is the method called residual kriging (also called detrended kriging). It consists of two components: one deterministic model and one stochastic residual model. The deterministic model is usually based on the relation between temperature and different topographical, land-use and other physiographical characteristics. The choice of parameters to include in the deterministic model depends on the regional climate conditions in the area to be mapped, access to external datasets to be used as predictors, etc. It has, however, become common to apply a multi-variate deterministic model including several explanatory parameters. Multiple linear regression is the most used method to establish the deterministic models. For long-term means of monthly, seasonal or annual temperatures, different deterministic models are established for each month or season [TVE 00], showing a seasonal variability which represents the different characteristics in the climate between seasons. If the study area is large the deterministic model might be regionalized, for example, by estimating the model parameters within a moving window [ENG 07], or smoothed between models developed for specific regions [DWD 99]. The atmospheric circulation will also affect the spatial distribution of temperature and this influence can also be identified in the deterministic model [UST 05].

The remaining residual field can be interpolated by applying any spatial interpolation method, like kriging, splines or inverse distance methods. Kriging is the most applied, and has the advantage to be based on a spatial structure function which is based on the stochastic theory and not on trigonometric or curve fitting techniques. By using spatial structure functions, a measure of the spatial uncertainty is provided.

The residual kriging approach is also applied on a daily scale, either by using the monthly deterministic models [TVE 05] or by performing a regression analysis on the daily datasets. The latter approach reveals large day-to-day variations in the coefficients of the deterministic model which can be related to the large scale atmospheric circulation [TVE 04].

6.7.2. *Spatialization of precipitation*

The spatial interpolation of precipitation is much more complicated than temperature. In contrast to temperature, precipitation is non-continuous in space and time. The spatialization of precipitation might be considered as two problems in one since both the occurrence and the amount of precipitation have to be assessed.

Due to a very high spatial variability, it is not an easy task to find a suitable spatialization method for precipitation. Different problems appear if we consider different time scales. The accuracy of the spatialization output also depends on the type of precipitation. Convective precipitation, especially the one that occurs during summer storms, is usually randomly distributed in space and could be limited over a narrow area, whereas stratiform precipitation is more evenly distributed in space.

Despite the importance of precipitation clear strategy of spatial interpolation of this element has been developed. However, for the estimation of long-term mean precipitation, the methods based on the principles of the PRISM-method [SCH 01] are widely accepted. This modeling approach takes advantage of different terrain characteristics such as slope, aspect, etc. into account. This is basically a linear regression approach, which enables the use of topographic information at several spatial scales. Another, rather simple approach is to apply triangulation with terrain adjustment [TVE 05]. This method has the advantage that it does not have a smoothing effect. It is, however, a pure interpolation method and the edges of the interpolated data will depend only on the location of the observation points.

Another approach that is frequently applied for the spatialization of precipitation is to apply the optimum interpolations concept which enables the use of first guess fields. These fields might be derived from numerical weather forecast models, or from remote sensing observations. A successful approach in this respect has been to combine radar information and *in-situ* observations. The mesoscale analysis system applied in Sweden, MESAN [HÄG 97] is based on this principle.

6.8. Climate indices

For many purposes, the end user does not need information about the basic meteorological or climatological elements like temperature or precipitation itself but rather information which is derived from these. In such environmental or societal applications there are gridded representations of weather elements or weather indices which are extremely useful. They could also serve as input to distributed models in other disciplines.

Many applications have focused on describing climatological seasons and so-called degree-days giving information, for example, of the length of the growing season, the accumulated degrees above or below a certain threshold describing this season, etc. (see, for example [TVE 01]). Especially for biological and ecological studies such information has proven to be valuable. Other indices which are easily adapted in a gridded climatology concept are traditional indices such as biocomfort and several drought and forest fire risk indices. This kind of GIS analysis forms a good basis for assessing the impact of climate change in different areas [SKA 04].

6.9. Gridded datasets

One natural result of spatialization is the development of gridded climatological datasets, representing time series of spatially distributed climate information at different temporal scales. By using all available information at any time the most reliable information should be established. This opportunity gives many possibilities for further applications and use of such information, especially for distributed modeling of environmental processes. However, there are some obstacles in the derivation of such datasets concerning the homogenity of the information. Observation networks are continuously changing over time and thereby the locations and the number of meteorological stations are also changing. On the left in Figure 6.4 a typical development in a station network is shown and thereby also the available information for spatial analysis. Another important issue is the quality of metadata. In order to maintain high reliability of the gridded data, precise information about the observations and the observation stations is needed. This problem is reflected on the right in Figure 6.4. It shows the development of the standard error of estimation in the monthly precipitation anomaly maps in Norway. For data prior to 1957, metadata is in some cases doubtful. This breaks down the spatial coherence, whose existence is the basic assumption of spatial modeling. It is therefore extremely important that the metadata has the same high quality as the observations used for spatialization.

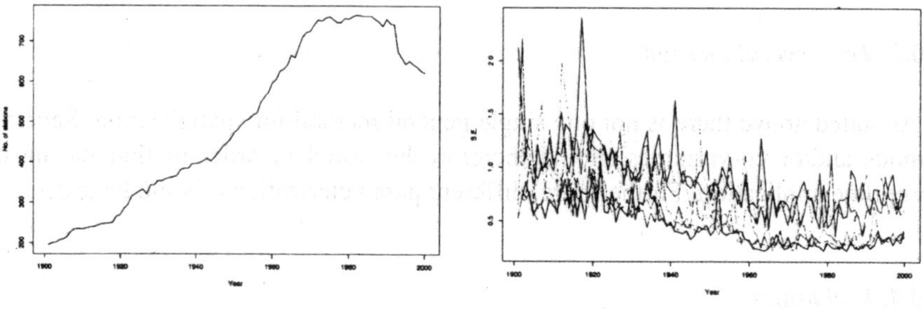

Figure 6.4. *The development of a typical station network (example of the precipitation network in Norway) is shown on the left. The standard error of a spatial interpolation of the monthly precipitation anomalies by cross-validation is shown on the right. Each line represents a calendar month*

6.10. Recommendations and future outlook

This chapter has briefly presented some of the development and status of the use of GIS in the spatialization of climate elements. COST 719 WG2 has monitored this development and collected examples of various spatialization approaches which are applied across Europe. Based on this activity a few general recommendations have been established.

6.10.1. *Choose the right method*

There is no general method that can be applied for any problem. The choice of method depends on the meteorological process that shall be analyzed, the available information, spatial and temporal scale. Another important issue is to consider the precision level needed by the final users, in order to select the most efficient method.

6.10.2. *Correct use of the method*

Most GIS software has some built-in spatialization functionality, making it easy to perform a spatial interpolation. However, it must be stressed that any use of spatial interpolation methods needs a certain knowledge of the assumptions and criteria which are associated with this methods. This also demands a detailed knowledge of the meteorological processes which is important in order to choose, for example, the best explanatory variables for fulfilling the demand of spatial stationarity.

6.10.3. *Test several methods*

As stated above there is not one single general method for spatialization. Several methods and/or co-variables should therefore be tested in order to find the most appropriate application. Furthermore, different parameterizations should be tested

6.10.4. *Validation*

The validation and assessment of uncertainty is possibly the most important task in spatialization. A proper validation is always needed in order to verify the results and to address the uncertainty of the spatialization.

6.10.5. *The future*

The spatialization of meteorological and meteorological elements has seen a tremendous development over the last decade. Accordingly, the need for spatially distributed information about weather and climate has similarly increased over the same period, mostly due to the high attention given to global warming and climate change by both scientists and policy makers.

The need for spatially distributed information about weather and climate will not decrease in the next years and the demand for such information with even higher spatial and temporal resolution will probably increase.

Spatialization within meteorology and climatology has still a large development potential and there are still many open issues to be solved. A recent trend within spatialization is the development of spatialization approaches which enable regional model parameters. Traditionally the method applied has one "global" parameterization, like one semi-variogram for the whole study domain, one expression describing the deterministic trend, etc. This in not likely to be the best approach and studies, also carried out within the COST719 framework, have indicated that regional or local parameterizations give better results. Methods like PRISM and MISH enable this, but also methods like residual kriging by applying a moving spatial window to determine the trend expression have proved to be efficient. The development of such approaches is in its early phase and it is expected that they will considerably improve in the near future.

Another similar approach that is under development is the use of weather-type information. It has been shown that atmospheric circulation has a strong impact on the spatial structure functions of weather elements and that such information can be used to improve spatially distributed estimates. This fact is one of the reasons to start a new COST action on the "Harmonization of weather types in Europe", COST733.

Numerical weather prediction models are getting an increasingly higher spatial resolution, and in many European countries they are now operational at spatial scales of 4-5 km. This means that they could provide more reliable first guess fields for the traditional spatialization methods. On the other hand, the spatialization methods presented in this report could also play a role as validation data for the NWP fields. The spatialization principles presented here will also be of great importance in the future for providing reliable spatially distributed climatological information.

A challenge that has to be solved when dealing with spatially distributed long-term climatologies is the problem of data homogenity. This is a topic that has been emphasized in traditional time series analysis of *in-situ* records. The issue of how to deal with the inhomogeneities of long-term climatology in a spatial context is not yet much discussed. This is an area for future studies.

6.11. Bibliography

[DOB 01] DOBESCH H., TVEITO O.E., BESSEMOULIN P. (2001) Geographical information systems in climatological application – Final report of project no. 5 in the framework of the climatological projects in the applications area of ECSN, DNMI Report 13/01 KLIMA.

[DWD 99] DWD, Klimaatlas der Bundesrepublik Deutschland, Teil 1, Offenbach am Main 1999.

[ENG 07] ENGEN-SKAUGEN T. and TVEITO O.E. (2007) Spatially distributed temperature lapse rate in fennoscandia, Proceedings of the Conference on Spatial Interpolation in the Climatology and Meteorology, Budapest, 24-29 October 2004, COST Office, in press.

[HÄG 97] HÄGGMARK L., IVARSSON K.-I., OLAFSSON P.-O. (1997) MESAN Mesoskalig analys, RMK 75 Report, SMHI, Norrköping, Sweden (in Swedish).

[JAN 07] JANSSON A., TVEITO O.E., PIRINEN P., SCHARLING M. (2007) NORDGRID Final report, met. no Report XX/2006 Climate, in prep.

[OBL 82] OBLED C. and CREUTIN J.D. (1982) Objective analyses and mapping techniques for rainfall fields – an objective comparison, Water Resour. Res., 18, pp. 413-431.

[SCH 01] SCHWARB M. (2001) The Alpine Precipitation Climate Evaluation of a High-Resolution Analysis Scheme Using Comprehensive Rain-Gauge Data. Zuercher Klimaschriften 80.

[SKA 04] SKAUGEN T.E. and TVEITO O.E. (2004) Growing-season and degree-day scenario in Norway for 2021-2050, Climate research, vol. 26:221-232.

[TVE 00] TVEITO O.E., FØRLAND E.J., HEINO R. HANSSEN-BAUER I., ALEXANDERSSON H., DAHLSTRÖM B., DREBS A., KERN-HANSEN C., JÓNSSON T., VAARBY-LAURSEN E. and WESTMAN Y. (2000) Nordic Temperature Maps, DNMI Klima 9/00 KLIMA, Oslo.

[TVE 01] TVEITO O.E., FØRLAND E.J., ALEXANDERSSON H., DREBS A., JONSSON T., TUOMENVIRTA H., VAARBY-LAURSEN E. (2001) Nordic Climate Maps, DNMI Report 06/01 KLIMA, Oslo.

[TVE 05] TVEITO O.E., BJØRDAL I., SKJELVÅG A.O. and AUNE B. (2005) A GIS-based agro-ecological decision system based on gridded climatology, Meteorological Applications, 12, 57-68.

[TVE 07] TVEITO O.E. (2007) Variations of daily mean temperatures and atmospheric circulation, Proceedings of the Conference on Spatial Interpolation in the Climatology and Meteorology, Budapest, 24-29 October 2004, COST Office, in press.

[UST 05] USTRNUL Z. and CZEKIERDA D. (2005) Application of GIS for the development of climatological air temperature maps: Meteorological Applications 12, 43-50.

Chapter 7

The Spatial Analysis of the Selected Meteorological Fields in the Example of Poland

7.1. Introduction

The application of GIS in climatology, and especially in meteorology, encounters many problems which are associated with integrating significant quantities of various types of information, as well as with taking into account the dynamics of phenomena which occur. These problems are particularly visible in the National Meteorological Services (NMS) which carry out operational tasks. The sources of such a situation are significant divergences in the data formats used in meteorology/climatology and in GIS tools, along with a lack of data converters/filters. The GIS systems have an insufficiently developed capability of manipulating the time dimension, which is very important for the rapidly changing meteorological data. There is also little interest on the part of GIS producers in an area which, in the initial period of the technology's development, did not promise significant revenues, but required costs to be incurred for the preparation or adjustment of standards to its specific requirements. In recent years, we have been observing positive changes in this area, such as the cooperation between the atmospheric sciences and GIS communities, which should contribute to the wider use of GIS tools in operational issues as well.

The functional capabilities of GIS are used to visualize data (e.g. ongoing measurement data or climatological characteristics), carry out spatial interpolation of

---

Chapter written by Izabela DYRAS and Zbigniew USTRNUL.

meteorological parameters, geometric correction of data (e.g. satellite images), pre-process data for mathematic models, integrate data from various sources and conduct analyses on them. Later surveys show that in recent years, changes have ensued in the direction of:

– A higher level of automation in spatial analyses.

– A more widespread use of GIS in "operational" issues.

– A more widespread use of freeware and shareware tools (especially in problems associated with spatialization).

There also appears to be an increased use of GIS in visualization as well as distribution of data and information.

The functional capabilities of GIS tools are an important factor, but not the only one for determining the use of these systems in meteorology and climatology. The access to geographic data is very important. As was presented in [MAD 05], the scope of geographic data use in meteorology and climatology can be considered for two aspects:

– Their use in analytical issues – that is, for various kinds of spatial interpolation and modeling – where the data is a factor that determines the spatial distribution of the meteorological phenomena or climatologic elements being analyzed.

– Their use in visualizing the results of measurements or calculations, in which case they are mainly used as background for other data.

The application of the GIS methods in climatology is still not very frequent in Poland although some progress has been made in the recent years when several projects adapting GIS to climate issues have been carried out. This is partially due to the lack of studies concerning spatialization methods. A poor choice of methodology can generate errors which result in maps that are totally inaccurate. Much better situations can be seen in other countries where in the recent years these maps constructed by means of GIS techniques have been fairly frequent (for example, see [TVE 00, DOB 01, BRO 02, TVE 02, COS 06]). Although most of the mentioned examples are based on the mean seasonal or monthly values, they present the basic climatic indicators showing the average mesoclimatic diversity and the general climatic conditions occurring on a given area. Moreover, the average characteristics of a chosen climatic element consist of many various weathers which may occasionally cause totally different meteorological conditions.

The main objective of this chapter is the presentation of different methods and tools for the spatial presentation of some meteorological/climatological elements. Simultaneously, special attention is paid to the validation of different interpolation methods for the construction of climatic maps.

7.2. Spatialization problems using standard observation data

GIS has become a very important and widespread tool serving a variety of functions in climatology issues (see [CHA 03]). Despite the significant progress that has been made in recent years, as regards the use of GIS methods and tools in meteorology and climatology, there are still problems to be solved. The most important of them is the question of interpolation or so-called spatialization in the process of recreating the spatial distribution of the respective fields. Although a number of new methods of interpolation have been developed, finding an optimum solution is still problematic and requires independent tests in each individual case.

Many spatial methods divided into several groups have been tested. The first group contains different kriging procedures such as: simple kriging, ordinary kriging, co-kriging, universal kriging and residual kriging. The second one uses deterministic methods: inverse distance weighting (IDW), splines and different trend surface analyses including regression approaches. Simultaneously, a couple of climatic elements have been tested (air temperature, precipitation totals, total cloudiness). The dataset contains mean monthly temperatures, mean monthly cloudiness and monthly precipitation totals from 168 stations (synoptic and climatological type) in the entire territory of Poland as well as from 55 stations located in the bordering zones (however, for temperature values only). Several geographic parameters including elevation, latitude, longitude and distance to the Baltic coast were used as predictor variables for air temperature interpolation. The first set of maps was constructed for the mean annual, seasonal and monthly temperature for Poland as well as smaller regions. For this element three methods (i.e. simple regression, multiple regression and residual kriging) generally gave the most reliable results (see Figure 7.1). However, almost all the validation parameters (ME, MSE, RMSE) confirmed that the residual kriging was the most relevant. Therefore, all these maps have been constructed with the application of the residual kriging. Special attention was paid to temperature parameters with practical value (e.g. the growing season length, duration of thermal summer and winter, and degree-days accumulations).

The analyses of the other elements such as precipitation and cloudiness did not bring such significant and unquestionable results. All the studies performed for the different spatial and temporal scales showed that the application of GIS techniques was a useful and promising tool for constructing climate maps. There are several spatialization methods which are suitable for the construction of climatic maps. However, residual kriging and regression methods seem to be the most adequate for the spatial interpolation of air temperature in Poland and probably in Central Europe. Residual kriging and IDW give the best results using the precipitation totals, whereas simple kriging is the most reliable for the total cloudiness spatialization.

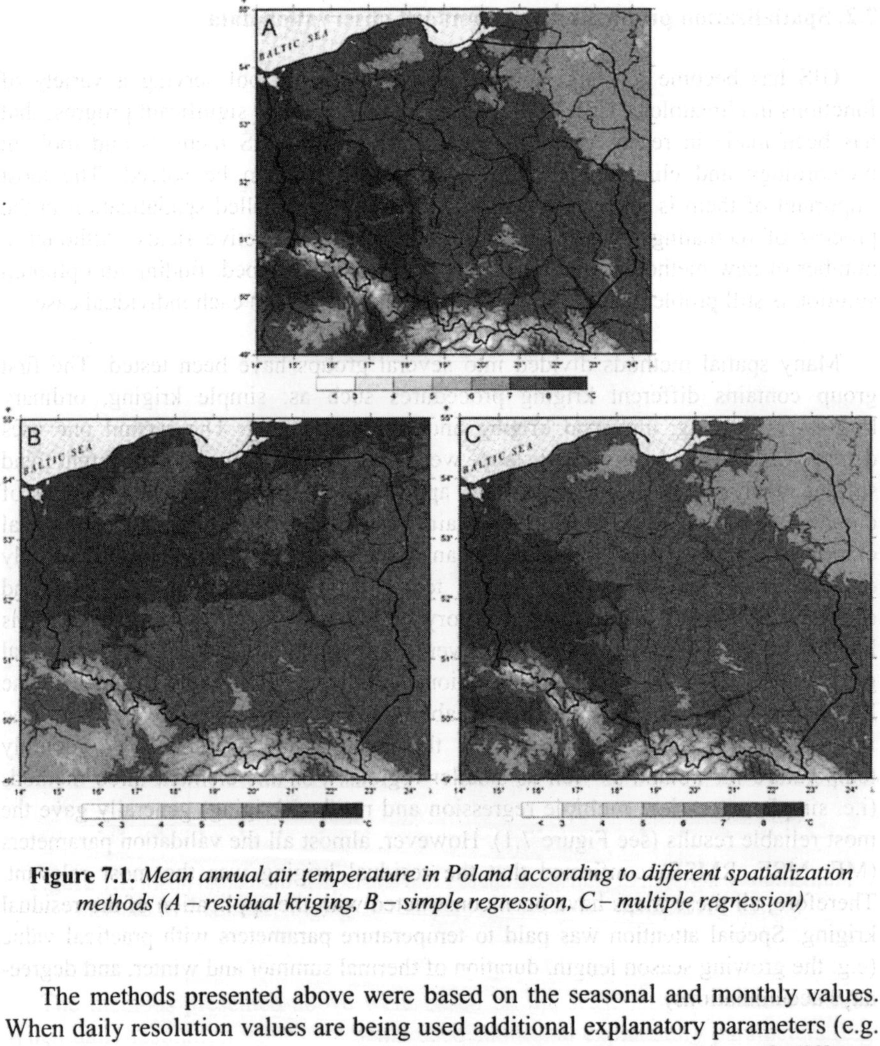

Figure 7.1. *Mean annual air temperature in Poland according to different spatialization methods (A – residual kriging, B – simple regression, C – multiple regression)*

The methods presented above were based on the seasonal and monthly values. When daily resolution values are being used additional explanatory parameters (e.g. circulation types) can be very helpful. There is a large variety of different classifications which can be used for different spatial applications (see [TVE 03]). For example, the application of the circulation types – as the main predictor – can be used for the spatialization of daily temperature in Poland or Germany (see [BIS 07, UST 05]). Their application usually improves spatialization results. Circulation types can help to obtain a more accurate spatialization, especially in winter. However, using some non-advective types does not improve results. Much worse results have been obtained for July when the application of circulation types for spatialization does not help as much as in January. This is connected with a weaker atmospheric circulation influence on the weather and particularly the temperature rather in summer than in winter. During the summer months the solar radiation predominates and therefore the role of the atmospheric circulation is not so important. Therefore, in the summer months the application of many circulation types cannot improve results and their implementation can be omitted.

7.3. Spatialization using remote sensing data

Meteorological and climatological applications of the remote sensed data have rapidly been developing in recent years. The robust spatial potential that the data obtained from meteorological satellites (polar-orbiting – NOAA – and geostationary – METEOSAT) constitute an invaluable source of information about the atmosphere and surface of the Earth. In particular, it provides information with reasonably good spatial and temporal resolution and therefore is a complimentary source of data in the operational and scientific practice. To make the most of satellite data, it is necessary to also use ground measured meteorological data which enables the creation and verification of algorithms for determining meteorological parameters from the satellite measurements. An analysis has been carried out for precipitation. Developed algorithms are in turn applied for determining the intensity and range of precipitation (see [DYR 05b]). These algorithms were prepared using NOAA (National Oceanic and Atmospheric Administration) satellite data. The GIS techniques presented later on will describe methods which enable the visualization of the joint analysis of satellite data and other meteorological data.

Satellites from the new NOAA generation are equipped with the advanced multi-channel radiometer microwave AMSU (Advanced Microwave Sounding Unit) for atmosphere sounding. This unit enables precipitable water distribution sounding and detection of the ice crystals in the atmosphere. This radiometer enables vertical humidity and temperature profiles retrieval in all weather conditions. The ER Mapper software was used to build the regression algorithms which enable the estimation of the total precipitable water and precipitation intensity from the satellite microwave data AMSU.

These algorithms were created as a result of the statistical analysis of the standard meteorological data (aerological sounding and synoptic data) correlated with the microwave sounding. The regression methods were used for the estimation of the state of the atmosphere over land (see [DYR 05a, DYR 05b, DYR 05c]). Figure 7.2 shows the example of the satellite image NOAA/AMSU dated 13.08.2004 at 12:27 GMT, with the distribution of the total precipitation probability (Crosby test). Each pixel is shown as a dot with the different grey shades corresponding to the probability value. The rain intensity contours are also displayed in various grey shades and are labeled. The IDW method was used for the spatialization of the rain rate in this case. The evaluation of the algorithm for the rain rate is positive. The areas with the high precipitable water content were in agreement with the aerological measurements and were characterized by a high convective activity. The regression algorithm for rain intensity has the tendency to decrease the high values and is better suited for moderate intensities (see [DYR 05a]).

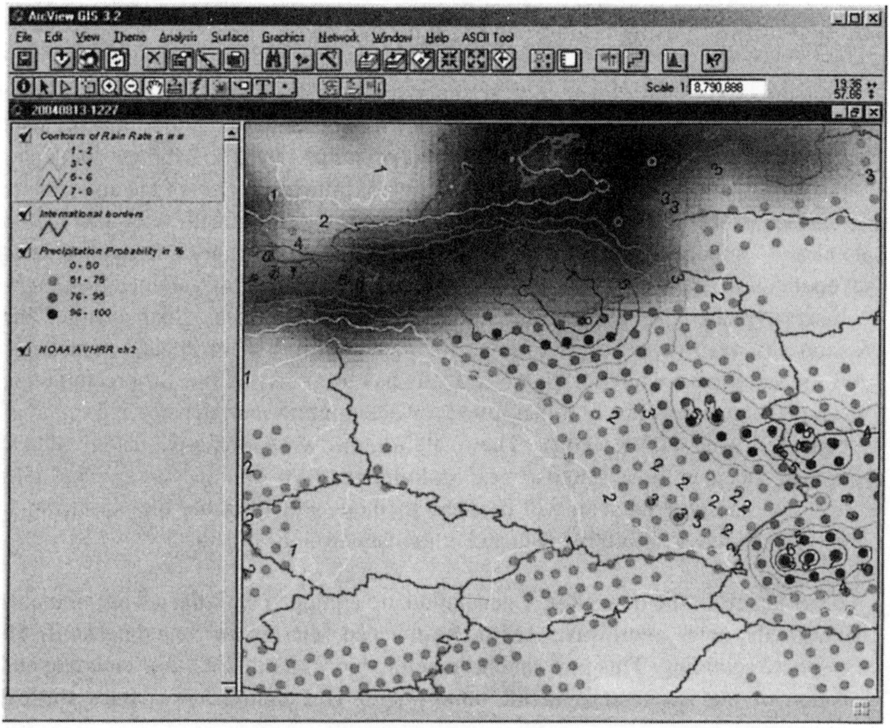

Figure 7.2. *The satellite image NOAA/AMSU dated 13.08.2004 at 12:27 GMT, showing the distribution of the precipitation probability together with the rain intensity contours*

The information regarding the humidity of the atmosphere detected from the microwave satellite data can be an important supplement for standard measurements. Aerological stations are located several hundreds kilometers apart while the satellite data has a resolution of 30 km. The NOAA revisit frequency is 6 times a day and therefore the satellite data can provide better information regarding the air masses' humidity. Presently the method works in experimental mode and the maps of rain intensity and range are distributed on the Intranet.

In the case of MSG (Meteosat Second Generation) two different products have recently been created. One of them is called PC (Precipitating Clouds), i.e. weak and moderate rain probability calculated from MSG/SEVIRI data. This class denotes the precipitation with the intensity 0.1 – 5.0 mm/h.

This algorithm is based on the linear combination of the visible and infrared SEVIRI channels and the surface temperature taken from NWP (Numerical Weather Prediction) data. Figure 7.3 shows PC for MSG data on the 16th December 2005 at 15:30 UTC. In these calculations the GRIB data from the mesoscale Aladin model was used.

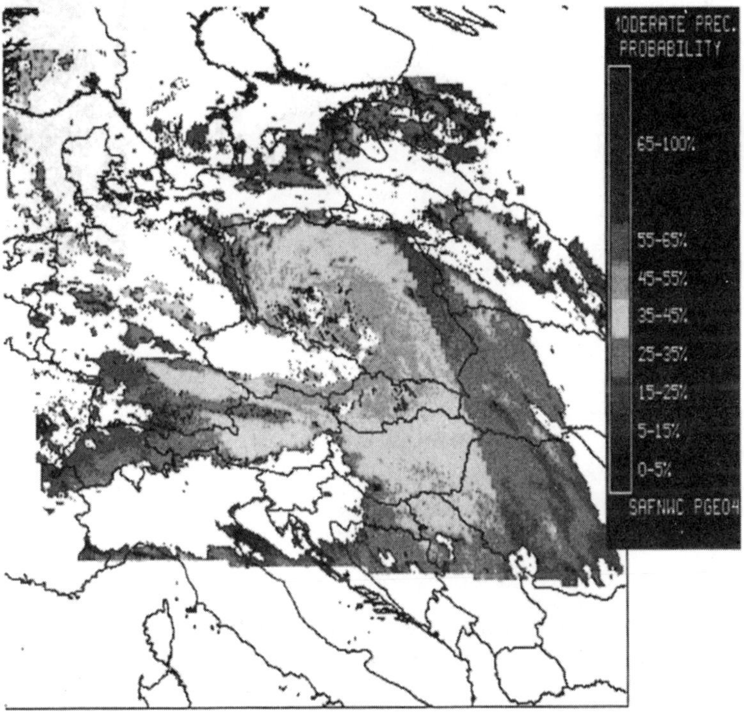

16 December 2006 15:30

Figure 7.3. *Probability of weak and moderate precipitation on the 16.12.2006 at 15.30 UTC in Poland. "Satellite" projection*

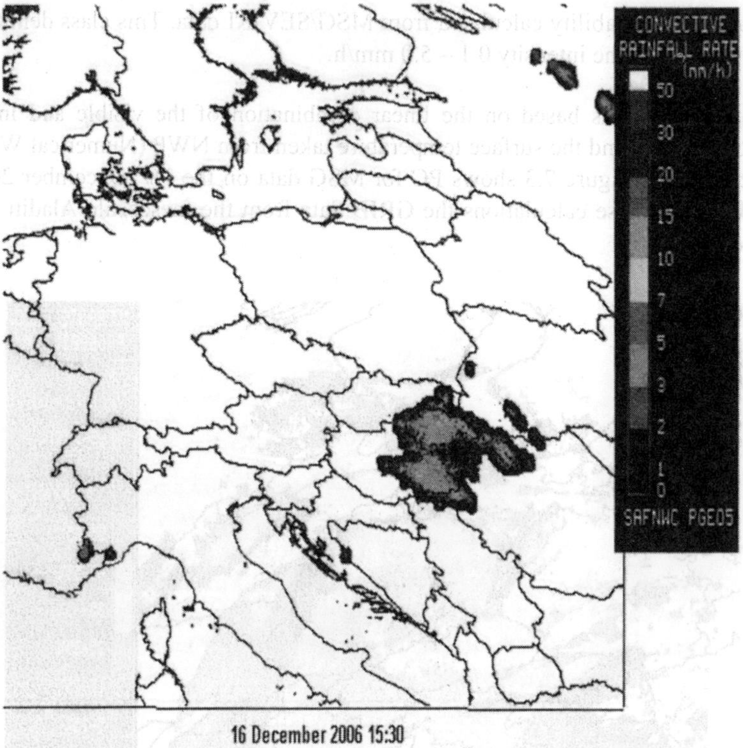

Figure 7.4. *Precipitation intensity from convective clouds, CRR on the 16.12.2006 at 15.30 UTC in Poland. "Satellite" projection*

The second algorithm is CRR (Convective Rainfall Rate) – i.e. convective rain intensity. Parameter CRR is calculated with the statistical method by using the correlation between multi-spectral satellite measurements and the concurrent precipitation which is measured by the radar observations. The rain intensity relation to the brightness temperature in channels IR, WV and albedo in VIS is denoted in a calibration matrix. Figure 7.4 shows the distribution of CRR for MSG data on the 16th December 2005 at 15:30 UTC.

Both products are calculated by the NWC SAF software provided by EUMETSAT and are obtained in HDF 5 format. The works are carried out in order to transfer the data into GIS software which makes it possible to compare NOAA as well as MSG products for further evaluation.

7.4. Conclusions

The preparation of meteorological/climatological maps is a complicated task. It requires the careful and detailed analysis of the respective element fields and the thorough knowledge of the physical processes connected with the complexity of the geographical environment. It is also difficult due to the fact that the standard data is representing the phenomena measured in points. Therefore, the application of remotely sensed data creates new possibilities and prospects for complimentary measurements with much better spatial and temporal resolution. In both cases special attention must be paid when choosing the most accurate and adequate spatialization method.

There is no simple universal spatialization method which is relevant to different climatological variables as well as for different spatial and temporal scales. Each element and resolution requires an individual approach. It was found that residual kriging is the best for monthly and seasonal means of air temperature and precipitation totals. The spatialization of the precipitation totals is particularly difficult due to its highly temporal and spatial differentiation. However, for daily and/or specific observational terms other methods could be used.

The validation method should consist of different indices, with at least a few of them depending on the type of element as well as the spatial and temporal variability.

The satellite data is helpful for precipitation spatialization. Satellite soundings of humidity and precipitation are necessary to supplement the information obtained from the traditional ground network. However, preliminary results combining both data sources are still far from fulfilling expectations.

Satellite data are quite robust for measuring water content in the atmosphere. Ground measurements have still high importance for calibration and validation of the remote sensing data.

The application of the circulation types as the main predictor usually improves the spatialization of the daily values for most climatic elements. However, using some non-advective types does not improve the results. Maybe the application of some additional, explanatory parameters should be considered (e.g. humidity, vertical profiles, air masses types, etc.).

7.5. Acknowledgements

This study was partly supported by a grant from the Polish Ministry of Science SPUB-M No. 618/E-217/SPB/COST/KN/DWM50/2005-2006. Figures 7.3 and 7.4

were included courtesy of Mrs Danuta Serafin-Rek from the Satellite Research Dept., IMWN, Krakow, Poland.

7.6. Bibliography

[BIS 07] BISSOLLI P., 2004, Using weather types classifications for climate maps, International Conference, Budapest, 169-175.

[BRO 02] BROWN D.P., COMRIE A.C., 2002, Spatial modeling of winter temperature and precipitation in Arizona and New Mexico, USA, Climate Research, Vol. 22, 115-128.

[CHA 03] CHAPMAN L., THORNES J.E., 2003, The use of geographical information systems in climatology and meteorology, Progress in Physical Geography, Vol. 27, No. 3, 313-330.

[COS 06] COST719 Final Report, 2006, Chapters 2 and 3: "Spatialization of the climatological and meteorological information by the support of GIS" and "GIS Applications", Brussels.

[DOB 01] DOBESCH H., TVEITO O.E., BESSEMOULIN P., 2001, Final Report Project no. 5 in the framework of the climatological projects in the application area of ECSN "Geographic Information Systems in Climatological Application", DNMI, KLIMA, Oslo.

[DYR 05a] DYRAS I., 2005, "Satellite Derived Precipitation Mapping Using GIS Technology"; Remote Sensing for Environmental Monitoring, GIS Applications, and Geology V, Proceedings of SPIE, Eds. M. Ehlers, I.U. Michel, Bruges, Belgium, 19-20 September 2005, Vol. 5983, 598313-1-5983-11.

[DYR 05b] DYRAS I., SERAFIN-REK D., "The application of GIS technology for precipitation mapping", 2005, Meteorological Applications, Special Issue on the Use of GIS in Climatology and Meteorology, Vol. 12, Issue 01, 69-75.

[DYR 05c] DYRAS I., ŁAPETA B., SERAFIN-REK D., 2005, "The retrieval of the atmospheric humidity parameters from NOAA/AMSU data for winter season", International TOVS Study Conference-XIV Proceedings, Beijing, China, 25-31 May 2005.

[MAD 05] MADEJ P., DYRAS I., USTRNUL Z., BARSZCZYŃSKA M., KUBACKA D., 2005, The use of GIS in meteorology, climatology and hydrology, Roczniki Geomatyki, PTIP Warszawa, vol. III, 3, 115-124, in Polish with English translation.

[TVE 00] Tveito O.E., Forland E.J., Heino R., Hanssen-Bauer I., Alexandersson H., Dahlstroem B., Drebs A., Kern-Hansen C., Jonsson T., Vaarby-Laursen E., Westmann Y., 2000, Nordic Temperature Maps, DNMI, KLIMA, No. 9.

[TVE 02] TVEITO O.E., SCHÖNER W. (eds.), 2002, Applications of spatial interpolation of climatological and meteorological elements by the use of geographical information systems (GIS), DNMI, KLIMA, No. 28, Oslo.

[TVE 03] TVEITO O.E., USTRNUL Z., 2003, A review of the use of large-scale atmospheric circulation classification in spatial climatology, DNMI, KLIMA, No. 10, Oslo.

[UST 05] USTRNUL Z., CZEKIERDA D., 2005, Application of GIS for the development of climatological air temperature maps: an example from Poland, Meteorological Applications, Special Issue on the use GIS in Climatology and Meteorology, Vol. 12, 43-50.

Chapter 8

Optimizing the Interpolation of Temperatures by GIS: A Space Analysis Approach

8.1. Limits of the interpolation in a heterogenous space

The process of interpolation is not a single, scientifically formalized method. Therefore, depending upon the method used, the results can be vastly different without being mathematically false. With the same dataset, it is possible to propose two very different maps (Figure 8.1).

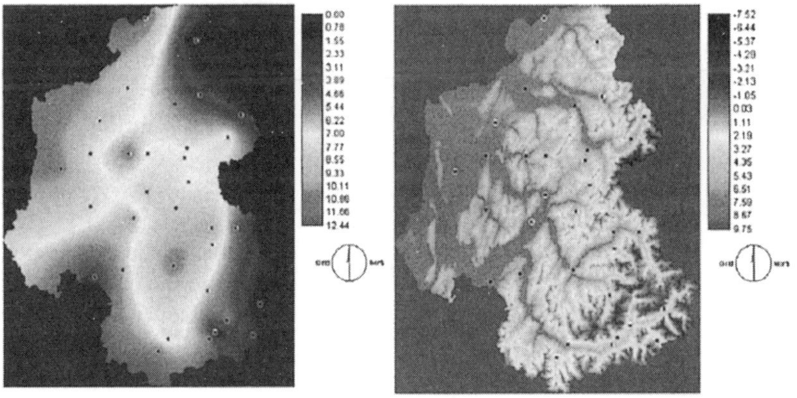

Figure 8.1. *Interpolation of the average temperatures in June for the Savoy and High Savoy departments. On the left, using kriging; on the right, using a linear altitude gradient*

Chapter written by Jean-Christophe LOUBIER.

The example shown here is a good illustration of this problem. By using the same dataset, on the left, we performed a kriging interpolation and, on the right, we followed a linear, altitudinal gradient-based interpolation. However, both models are validated mathematically on the basis of both the general macro-structural appearance and the local predictions and they each result in dramatically different models of temperature. One of the most striking examples of the resulting discrepancies can be seen at the top of Mont Blanc: the linear model proposes a value of –7°C whereas the probabilistic model estimates the value at 7°C. More than 14°C of variation!

The reason for such large differences can be traced back to the mechanism of interpolation. Interpolation is simply a mathematical "recipe" that applies to the field as a whole. If the region in question is morphologically homogenous, then the process can be applied with good reliability. However, if the space to be treated is quite heterogenous and the web of weather stations does not effectively represent this heterogenity, it becomes impossible to obtain a valid representation for the field without technical input from an expert.

8.2. Optimizing the spatial distribution of the stations

The quality of the interpolation is dependent upon both the number of measured points present in a given field and on the way in which these points are laid out. The more complete the sowing, the less the error will be significant, as the number of spaces without information becomes limited. However, since we cannot always choose the number or position of weather stations, we propose a method to optimize the sowing of weather stations through the creation of virtual stations. In this way, we "mechanically" reduce the errors of interpolation. We can then hope to reach space estimates of good quality starting from an irregular or sparse grid.

8.3. Underlying space assumptions

The surface to be interpolated has particular space characteristics, corresponding to the phenomenon measured in space and time. According to the basic paradigm of space analysis two spatially close points are more similar than two distant points. For our project, we also assume that two points sharing similar physical attributes (altitude, orientation, etc.) are more similar than points with disparate characteristics.

8.4. Theoretical structure of our model

At the beginning, we have a series of points that are temperature measurements on a map (X; Y). Our goal is "to guess" the temperature at an unknown point on the map (that we will name P). If we assume that temperature is a phenomenon which is determined by a certain space and by certain climatic conditions, then we can propose that the value of the temperature at point P is a known average temperature modified by a value, wi. The latter is a function of the physical phenomena that affect the temperature at point P, for example, the distance between P and the other points on the map. The equation for this approach can be written: T (Xp) = T (Xi), where Xi is the value at point A or B or….N and Xp is the value at the unknown point, with T being, of course, T°. The developed form gives:

$$T(Xp) = waT(Xa) + wbT(Xb) + ... + .wnT(Xn)$$

with $\sum_{I} wi = 1$ because the totality of the wi corresponds to 100% of the studied phenomenon.

The difficulty is in establishing the stabilizing coefficient wi using the spatial properties that specifically influence the value at point P. For this chapter, we will use: altitude (alt), orientation/exposure (the East) and the distance to a point of comparison (Xp). However, these factors are not the only possible ones. We can add any element that we judge relevant in the determination of the predicted phenomenon as long as it is possible to mathematically model these elements under a linear constraint.

8.4.1. *Information management in GIS*

A certain amount of information concerning the known and unknown points of our field is given in the Geographic Information System (GIS) database (Figure 8.2). For the known points we know: the temperature, the orientation, the altitude and the distance between the measurement points. For the unknown points, we would like to calculate T° but we know the three other parameters from the GIS database. As we have $\Sigma = 1$wi, we can build an interval ranging from 0 to 1 to linearly model the effects of the chosen factors on wi for each "virtual station". This helps us to model these influential factors on the temperature, since the functions chosen to model the phenomenon can only result in one solution for any point on the resolved map.

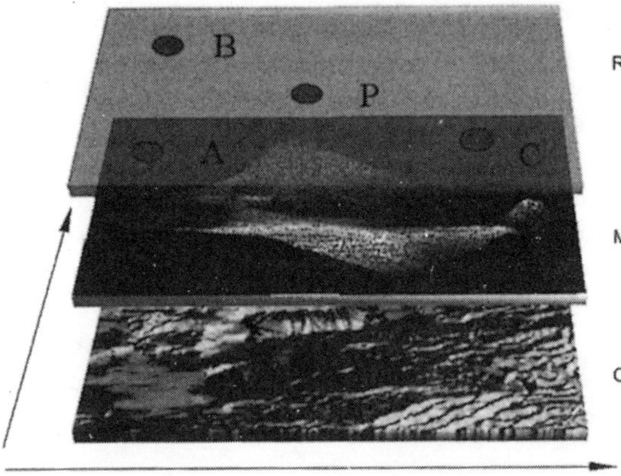

Figure 8.2. *Data management in the GIS*

8.5. The process of linear modeling for the selected factors

Altitude generally effects temperature in a linear way [OZE 86, CAR 82]. It is reasonable to think that two points at similar altitudes will have similar temperatures. Thus:

– if A (Xp) – A (Xi) = 1 then the similarity is maximal

– if A (Xp) – A (Xi) = 0 then the similarity is minimal

with A as the value of wi.

We can thus model the effect of altitude by using a linear equation where the value read on the y-axis, for the known point X, will correspond to the contribution of the real point on the unknown point P. The slope of the function is given by the two points with coordinates (0; 1) and (alti max-alti min; 0). For orientation, the model includes an additional step but it is based on the same principles. We must first take all the rough values for orientation in degrees through an interval ranging between [–180 > 0 < 180]. The selected equation will thus invariably take the form of an interval given by the pair of items (0; 1) and (180; 1–0). In order to model distance, we use the same method once again. This function develops into the interval (0 max), (Dist Xi Xp Dist of the field).

8.6. Determination of the optimal positioning of P

Now, since we have the possibility of building a new point on our map and determining its T°, the question turns to where we can add a virtual station to improve the sowing of temperature stations. Rather than taking a systematic approach (with a regular grid on the field), we have chosen an iterative approach in order to attempt to find the temperature at spatially interesting points. For example, we could choose points which are distant from all other known points and then successively choose N points until the map coverage is optimal (when both the model confidence and the error from cross-validation are minimal). This iterative method makes it possible to optimize the computing time and only leaves the judging of the quality of the interpolation to the expert.

8.7. An example of implementation

We tested this model on a panel of 33 weather stations extracted from base ER30. This data was collected by Research Team no. 30 (Cartographic Research applied to the Climate and the Hydrology of CNRS) over a 25-year period spanning from 1967 to 1992. The behavior of our model was subjected to cross-validation. For this, we removed individual weather stations from our data file of temperatures and we reintroduced them by using only their spatial characteristics. Afterwards, we compared the two matrices of results (real and envisaged). This analysis shows a stable behavior over time, since we obtain a determination coefficient of 0.76 between the two matrices. This means that the seasons are well represented by our model, an observation that is confirmed by the study of the tôtal monthly averages (Figure 8.3).

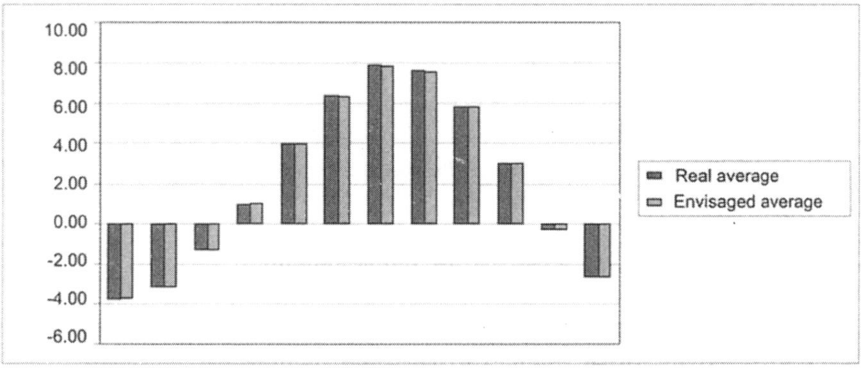

Figure 8.3. *Comparison of the real and envisaged monthly averages*

On the level of individual stations, however, the model yields disparate results. Certain weather stations are well estimated by the model (Figure 8.4), whereas others are badly represented (Figure 8.5). The latter in particular are all thermal situations compared to the distribution as a whole. This means that the virtual station model fluctuates around the average and the average plays an attractive role.

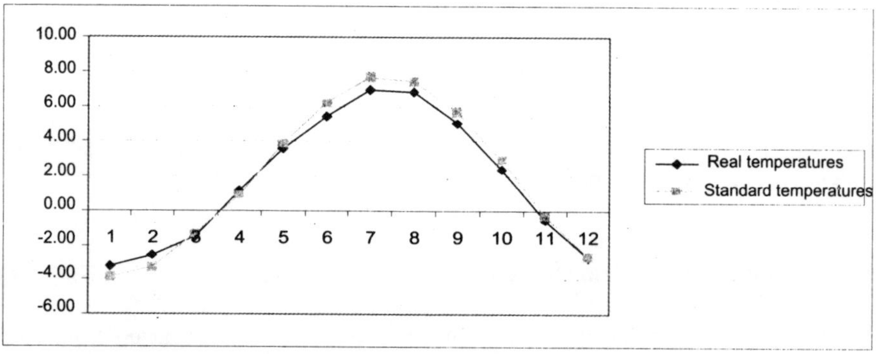

Figure 8.4. *Type of station with good quality forecast (Avrieux)*

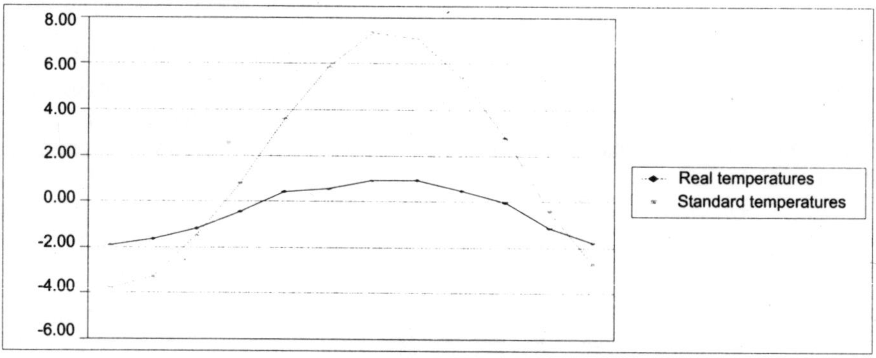

Figure 8.5. *Type of station with bad quality forecast (Bessans)*

8.8. Consequences and spatial/structural understanding

The fact that this model provides a calculated temperature that is dependent on the position of the virtual station is interesting information. By performing a systematic comparison of virtual stations with real stations, we can build a catalog of spatial situations where an interpolation will be of high or low confidence. According to the principle of space resemblance, which is used as a postulate in our model, it is possible to identify these spaces by using GIS raster analysis to find zones which are identical to those in our catalog. Thereafter, we could either add real stations in spaces that are spatially interesting (i.e. those that are the least well represented in the grid), or simply improve the sowing by the addition of virtual stations. In both cases, the optimization makes it possible to improve the quality of the final model of interpolation. The GIS approach that appears the most adapted for this exercise seems to be analysis by fuzzy logic.

8.9. Determination of authorized spaces

Our objective is to determine cells (pixels) that contain one or more specific spatial parameters in order to optimize the positioning of a virtual measuring site. In this context, two types of space criteria influence our choice: constraints and factors.

8.9.1. *Constraints*

These constraints are of a Boolean nature (0; 1) and they express the spatial preconditions necessary to satisfy our project. For example, altitudes and orientations which are not represented in our weather station data file are constraints because it is impossible for us to extrapolate beyond the limits of our dataset.

8.9.2. *Factors*

The factors make it possible to account for continuous phenomena in the analysis. It is a question of evaluating a criterion according to a gradient. The inter-station distance, the altitude and the orientation are included in this category. For example, the altitude 1,215 can be regarded as relevant because it corresponds to a station with a thermal profile which is well translated by the virtual weather station model. It is probable that the altitude 1,214 will be also acceptable, with an error generated by the drift from the optimum (altitude 1,215).

Fuzzy logic makes it possible to mathematically implement these factors [ZAD 65]. This theory is founded on the determination of a function of membership. The shape of this function (from 0 to 1) enables us to give a value indicating the degree of correspondence of each cell with a continuous spatial phenomenon (Figure 8.6).

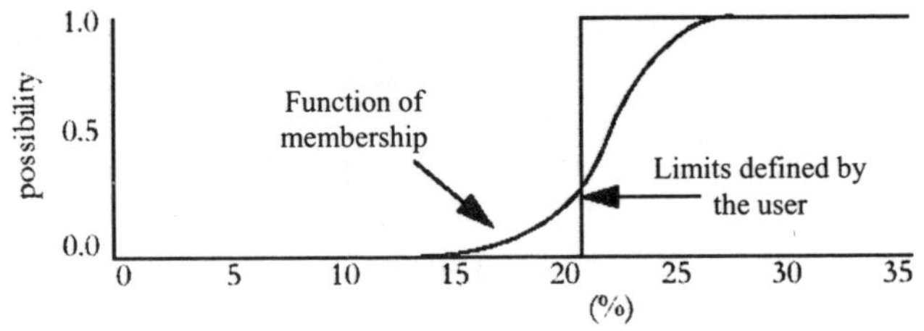

Figures 8.6. *The model*

For example, the use of a sigmoid function for the determination of the membership of all the values of altitude will give a continuous interval in which the altitudes are well managed by our space forecasting model. The critical point for defining membership is at the moment the function is created. It must represent a concept, for example, valid altitudes or invalid altitudes and is a choice of the user that must be justified.

8.10. Taking uncertainty into account: a choice/given couple

A number of factors influence the level of uncertainty in our model. The most obvious form of this uncertainty appears during the assignment of values to the points of inflection on the curve chosen by the user for the construction of the factors. During this process, a limited choice of values is related to the quality of the data available. Therefore, the error can be related to these two uncertainties: the quality of the data and the rules of the decision defined by the user.

For our analysis, we will seek to evaluate the error according to a couple confidence/assumption. The confidence part of this couple can be summarized by the following sentence: *the space information relating to our weather stations, which are well returned by the model, is the explanation of the temperature measured at this point*. The second part, i.e. the assumption, is defined in the following way: *the quality of the temperature envisioned at points with the equivalent space characteristics will be as good as of the points equipped with a real station*. This couple confidence/assumption forms a rule of decision of the postulate type. Bayes' law enables us to evaluate the probability of our assumption. Consequently, the final space result will be a selection of the best spaces according to the criteria retained on the assumption of probability.

8.11. The standardization process

The standardization is rather simple since it follows a continuous law. It is simply a question of operating a linear calibration in a common interval [0 and 255]. This calibration is established starting from the minimal and maximum limits (i.e. 0% of membership and 100% of membership).

Two strategies are possible. The first is based on the assumption of equiprobability, in which case we must consider that each factor exerts an equivalent weight in the process of identifying optimal spaces for the virtual station positioning. The second is based on weighting the space factors, as done in the mathematical modeling. It is this second method that was chosen, as it seemed logical to preserve the characteristics of our calibration (altitude = 0.4; orientation = 0.3; outdistance = 0.3).

8.12. Results for the addition of stations

The result of our analysis shows that we can create a rather significant number of virtual stations and improve the initial sowing of stations which were offered for interpolation (Figure 8.7).

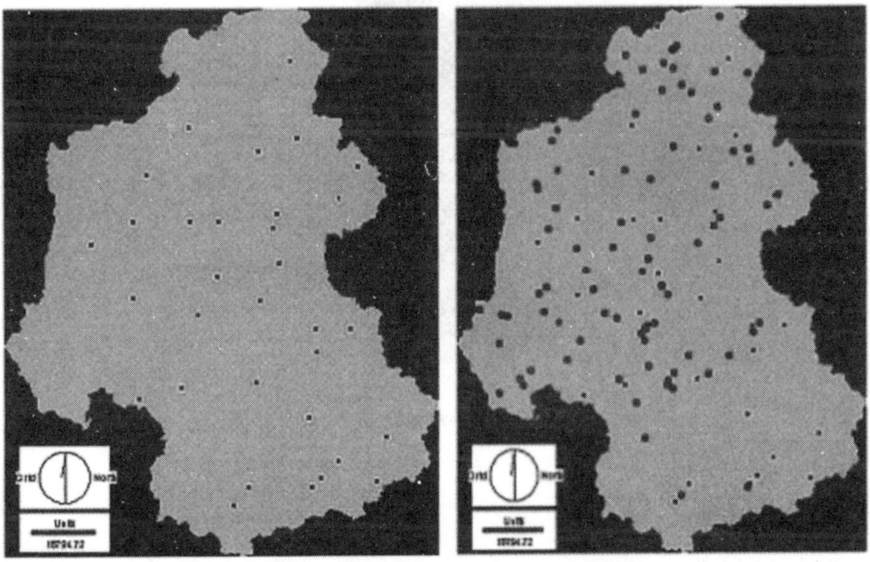

Figure 8.7. *Comparison between the initial sowing (on the left) and the improved sowing (on the right)*

8.13. Authorized interpolators

The use of this method implies a particular choice for the methods of interpolation of the web. The steps that we propose prohibit the recourse to probabilistic approaches like kriging. In our model, the position of a station on a map determines its temperature, whereas probabilistic methods, like kriging, insist that the measured values do not depend on the position of the points, but only on their distance [GRAT 02]. For this reason, the procedure outlined above can only be accompanied by deterministic interpolations. Among those, the bicubic spline of the radial basis function family gave very interesting results.

The example below represents the result for the average temperatures in January in the two departments, Savoy and High Savoy (Figure 8.8). The study of the residues in cross-validation (residues means 0.25 and variance 11.49) shows that this model is without skew.

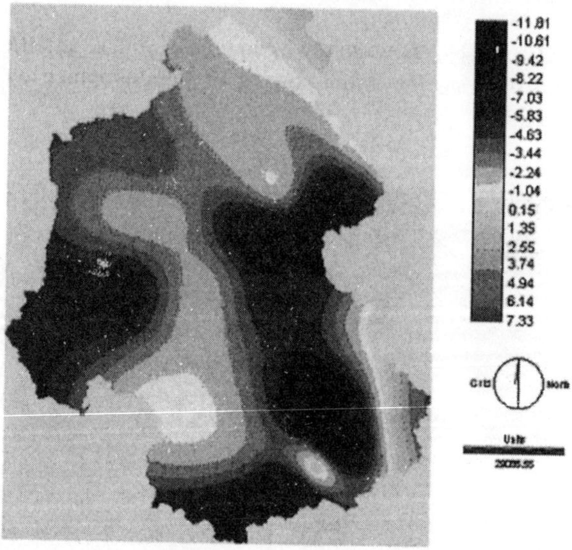

Figure 8.8. *Interpolation map by bicubic spline*

8.14. Conclusion

The method that we propose represents a practical and operational procedure to use when we need to improve the quality of interpolation for temperatures by using a loose, sparse sowing. From a more formal point of view, this approach shows that the process of interpolation is never equivalent in space. It appears that spaces are excluded from the procedure due to a lack of representation in the initial sowing. In this context, an interesting conclusion that we can draw is the fact that the implementation of a total method on the unit of a field masks site errors which are related to the configuration of the ground in a diffuse way. The analysis of the residues does not seem efficient enough to detect this type of error, except when implemented on a set of stations that were not used in the interpolation process.

8.15. Bibliography

[ARM 97] ARMSTRONG M. and CARIGNAN J.: 1997. Géostatistique linéaire: application au domaine minier: Ecole des Mines; Paris.

[ARN 00] ARNAUD M. and EMERY X.: 2000. Estimation et Interpolation spatiale: méthodes déterministes et méthodes géostatistiques: Hermes; Paris.

[CAR 82] CARREGA P.: 1982. Les facteurs climatiques limitants dans le sud des alpes occidentals; Thesis; Nice.

[CRE 79] CREUTIN J.D.: 1979. Méthodes d'interpolation optimale de champs hydrométéorologiques: Comparaison et application à une série d'épisodes pluvieux Cévenols; Thesis; INPG; Grenoble.

[GRA 02] GRATTON Y.: 2002. Le krigeage: la méthode optimale d'interpolation spatiale; Institut national de la recherche scientifique; Quebec; Canada.

[LAB 00] LABORDE J.P.: 2000. Méthodes d'interpolations et géostatistiques pour la cartographie automatique: CNRS UMR 5651 "Espace"; Equipe Gestion et Valorisation de l'environnement; Nice; Sophia-Antipolis.

[LOU 04] LOUBIER J.-C.: 2004, Perception et simulation des effets du changement climatique sur l'économie du ski et la biodiversité (Savoie et Haute-Savoie), Thesis; Joseph Fourier-Grenoble I University; 6 May 2004 available at http://tel.ccsd.cnrs.fr/documents/archives0/00/00/69/90/index_fr.html.

[OZE 85] OZENDA P.: 1985. La végétation de la chaîne alpine dans l'espace montagnard Européen: Masson; Paris.

[TON 95] TONG-TONG J.R: 1995. La logique floue: Hermes; Paris.

[ZAD 65] ZADEH L.: 1965. Fuzzy Sets: Information and control; Vol. 8; pp. 338-353.

Chapter 9

Daily Winter Air Temperature Mapping in Mountainous Areas

9.1. Introduction

Climatic databases were constituted, historically, on specific measurements established by people (with different methods: climatic parameters in weather station for current values or those of the recent past, dendrochronology or coring for estimates of older or historical temperatures).

These measurements are irregularly scattered data, so it is not always possible to include them into climate models or other applications: it is necessary to interpolate them.

Many official documents (published by national or international organizations) propose maps of temperature or precipitation – they are the most largely diffused climatic variables because they are easiest to measure and the most understood by the public – either as point data and therefore with a majority of vacuums on the map, or interpolated with spatial estimators (without taking into account the relief). These kinds of maps are mostly not very useful, which is why we need valid fields of climatic data and not point ones.

Chapter written by Rémi LHOTELLIER.

The difficulty to estimate these fields increases in an area with alpine relief, since it is obvious that in a mountainous zone the variations of relief are the first climatic factor influencing air temperature.

9.2. GIS and climatic data

Interpolating means intercalating values in a series of existing and known values. So interpolating temperatures, starting from punctually measured data, consists of realizing fields of temperatures. For this reason, the advantage of spatial climatic databases lies in their capacity at being integrated in GIS [CHA 03], even if, often, their sources are different. This makes it possible to carry out environmental studies at various scales. We try to determine, for a mountainous area, the influence on temperature of geographical factors under various weather types, as well as their scales of influence and their reliability for a spatialization of daily minimum and maximum temperatures at the local scale in the French Alps (about 24,000 km² surface, resolution 150 meters by pixel). Our studied area includes four French departments, from Lake Leman in the North to Durance in the South and from the Rhone valley in the West to the Italian and Swiss borders in the East.

The main interest of a work at the local scale consists of taking into account the local factors influencing the climate and their integration into temperature maps. For example, a map of maximum temperature for one winter day was selected below as an illustration because it underlines the effects of relief (as seen on Figure 9.5). Thus, in this figure, two maps present the maximum temperature on the 4th December. The exposure to solar radiation acts as the first cause of heated air temperature: south-facing slopes have values from 2°C to 4°C more than shaded thalweg. The north/south-facing slopes opposition involves variations of about ± 4°C in interior valleys. The amount of solar radiation is therefore a dominating element which governs thermal process. In this example, we are interested only in the theoretical shade induced by the relief and not in nebulosity. The digital elevation model of the northern part of the French Alps is needed, as well as the calculated slope and aspect.

The calculation of the potential solar radiation methodology is based on Van Dam' works [VAN 01]. The growth of radiation values during the year is quite obvious. Moutainous relief is responsible for the strong fluctuations of the calculated values on little surfaces.

The oppositions between slopes are all the more significant at local scale: Figure 9.1 illustrates the potential solar radiation received in a much contrasted surface. South-facing slopes can receive more than three times the energy measured for north exposures (6th March): in March, the sun has already begun its rise on the horizon but it is not yet high enough to strongly heat north-facing slopes.

This information is then included in a stepwise regression model with other topographic parameters to estimate air temperature. It might also be used for hydrological models or to estimate evapotranspiration.

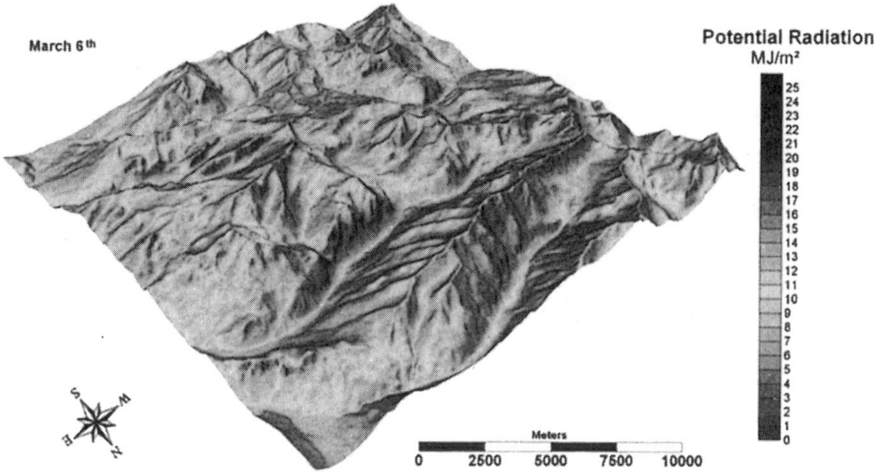

Figure 9.1. *Potential solar radiation estimated by GIS on 6th March*

Extreme daily temperatures are difficult to model because a minimum (or a maximum) temperature is not reached in every point of an area at the same time and also because the temperature variation factors are all the more numerous as the studied scale is local. For a very small area, land occupancy is, for example, a good estimator [ESC 81], but our studied scale does not make it possible to reveal the influence of this parameter.

The daily scale distribution of the part of the topographic variables is complex. Elevation generally remains the first explanatory factor, but other parameters bring a great part of variance. Thus, the influences of these parameters are connected to the atmospheric circulation. With this temporal scale appear local parameters (like curvature), although most of the time their contribution is disappointing.

9.3. Spatialization of air temperature on a daily scale

The datasets are provided by Meteo France (daily minimum and maximum temperatures from 1990 to 1995 in 168 weather stations). The method used is stepwise regression and residual kriging with a temporal scale varying from daily to yearly values. We focus on a daily scale in this example. The evolutions of statistical results according to the temporal scale (season, weather type and monthly/daily data) were exposed in other works [LHO 03, LHO 04].

The interpolation method used to spatialize temperature with residual kriging is founded on predictors (independent variables), especially topographic and environmental ones:

– elevation measured at the station;

– slope at the station;

– aspect at the station (projected with sine and cosine, or combined with slope);

– tangential, plan and profile curvature at the station;

– latitude and longitude of the station;

– potential solar radiation at the station;

– other factors describing local relief (like interfluve distance).

These predictors can be calculated in moving windows (low pass filters, for example) for a better estimate and some parameters can be improved according to the temporal scale. Weather circulation can also be included to improve the accuracy of the estimated values.

A Short-Term Scientific Mission in the COST719 framework [LHO 05b] was proposed to improve the spatialization of temperature, even in winter when temperature inversions are frequent. The first point was the creation of new topographic predictors to improve already obtained results. Some grids were tested, like relief shading, North-South and West-East modified aspect, integrated slope and aspect in one unique variable (slope-aspect factor), digital model of riverbeds' elevation, relief energy, distance between main flow paths and ridges, elevation range in moving windows, elevation standard deviation in moving windows.

Apart from these variables, we further tested the possibility of improving the regression models by:

– using as predictors the mean altitude values calculated within moving windows of different sizes (from 150 m x 150 m up to 24,600 m x 24,600 m);

– condensing the predictors' space by principal components analysis and using the first components within the regression models;

– applying the Aurelhy method [BEN 87] and using the first 15 EOFs for the regression models.

Moving windows centered on climatic stations (3 x 3, 5 x 5, 7 x 7 pixels, etc.) were created to take into account local and regional aspects of the terrain and their influence on temperatures. Filters were tested on several predictors (elevation, slope, tangential, profile and plan curvature, potential radiation, modified aspect) with significantly better results.

An enlarged window on elevation in winter may take into account temperature inversion in the regression model, but smaller windows are sufficient in summer. Therefore, using filters on elevation grid seems to be very useful to estimate winter temperatures. In fact, winter temperatures are often the most difficult to spatialize because of high pressure areas remaining on the Alps for days. Thus, generally, if a method works on winter temperatures, it will also function with summer temperatures (which are more correlated to elevation).

The equation used can be written as:

$$y(s) = \beta + \beta 1x1(s) + \ldots + \beta nxn(s) + \varepsilon(s) \qquad [9.1]$$

with

y(s): dependent variable

$x_1(s)$ to $x_n(s)$: independent or explanatory variables

β: coefficients

ε(s): residuals of the regression (difference with the model).

Then we interpolate ε(s) by kriging and sum regression grid with interpolation grid [TVE 00, TVE 02].

When explanatory parameters used by multi-variate regression are strongly correlated with temperature, the residual kriging only brings a weak additional explanatory share. On the other hand, its use is interesting when taking into account regional climatic effects. Moreover, it makes it possible to remove errors in estimation near the measuring sites with particular thermal behavior.

The best results are provided by filtered grids: many error values are near zero when filtered grids are used. The mean error is lower with filtered data than with point data. Standard deviation is also lower with filtered grids. Therefore, multi-variate stepwise regression and residual kriging give better results and more credible maps. One conclusion is that topographic variables must be integrated in moving windows, so they need to be filtered for winter days (useless for summer or annual scale).

The model was created using one sample (68 meteorological stations) and the validation was carried out on a second sample (69 others stations not included in the model) so we can compare estimated temperatures with observed temperatures (error = observed temperature – estimated temperature). For December 1995, 2,140 estimates are compared (Figure 9.2).

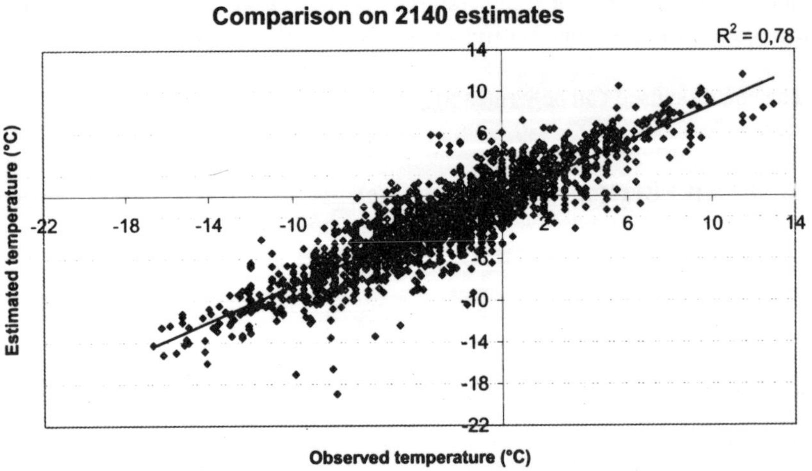

Figure 9.2. *Observed versus estimated daily air temperature in December 1995*

The validation of the method was tested on the daily minimum and maximum temperatures in December 1995. Results are presented here for the minimum temperature only, for 31 days. We compare the estimated temperature and the actual temperature in Figure 9.2. The estimated temperature is not always exact, but globally the results are correct. The biggest errors only come from some weather stations, which have singular thermal behavior and the sample is not sufficient to model their profile. The coefficient of determination (part of the explained variance) between estimated and measured temperatures reaches 78%.

Due to the time scale, we must focus on a few representative days and weather stations. Three weather stations are chosen to illustrate the validation: Mens (780 m), Modane (valley, 1,120 m) and Saint-Bernard (890 m). The three graphs show three different behaviors: air temperature is over-estimated at Saint-Bernard, there is an undervaluation at Mens and the estimate is quite good at Modane. We can conclude that factors involving a good estimate are:

– a sample which is broadly representative of the terrain;

– the proximity of some points of this sample to the estimated station;

– similar topographic characteristics between neighboring stations (necessary because of interpolation residuals).

As an example, Modane is surrounded by other valley weather stations so local effects are well included in the model. On the contrary, Mens and Saint-Bernard are far away from the others and using residual kriging may in this case worsen the results. The results are opposite in these two last stations because of local topographic conditions. Of course, the larger and more representative the sample is, the better the results of the estimate will be.

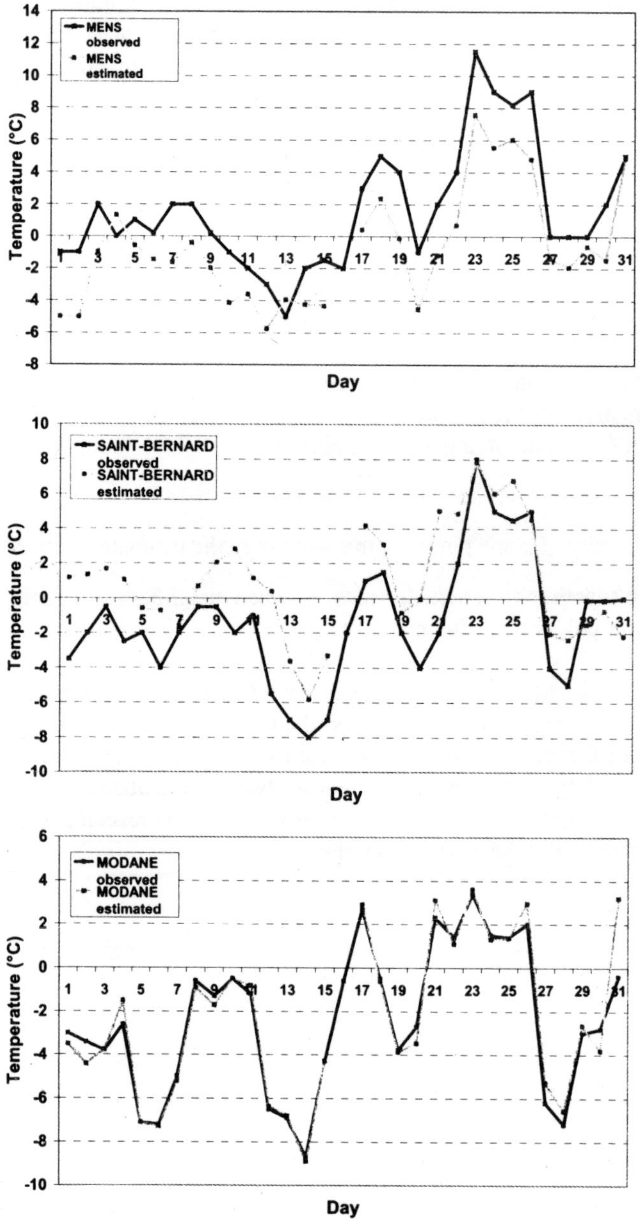

Figure 9.3. *Comparison between observed and estimated air temperature for the 31 days of December 1995, at Mens, Saint-Bernard and Modane*

9.4. Temperature maps (local scale)

The obtained maps are, of course, much more attractive than punctual maps. In Figure 9.4 (minimum temperature near Grenoble on 4 December 1995), we note a temperature inversion: the air temperature is cooler at the bottom of the valley than on the near sides. Nevertheless, above a certain limit (for this day, approximately 1,000 m), the temperature gradient (vertical lapse rate) approaches the adiabatic state.

3rd December 1995 (Figure 9.5) was chosen as an illustration because it is one of the rare days with a strong influence of the solar radiation. In fact, this influence is obvious in the distribution of air temperature but, because of the sampling and the situation of the measurement stations, it is very difficult to take it into account. Therefore, on this day, we can note a strong heating of the air with the effect of solar radiation: the effect of north-south exposure is very visible on the selected illustration.

All these maps propose a good representation of reality; nevertheless, uncertainties obviously remain in this spatialization.

The quality of the regression model, as well as the validity of the map, depends primarily on season and weather types. Unfortunately, a stepwise regression model does not function for all days. Regressions with a coefficient of determination from 10 to 30% are not useful because a simple kriging would provide similar results. On average, winter high pressures are responsible for the worst results because of temperature inversions. However, weather circulation is not always responsible for bad results (a weather-type classification at local scale, and not a circulation one, would be necessary for our daily scale) [TVE 03].

We stress that the statistical validity of the map must authorize whether it should be created or not: indeed, mapping temperatures remains possible even with a R^2 reaching 5%, but what is then the significance of the obtained map? The interpolated map must thus be accompanied by a validation map so that users can estimate its uncertainty. The information given by the coefficient of determination or cross-validation, for example, is essential. We can also imagine validity maps [LHO 05a].

Numerous maps and indicators can be deduced from temperature [UST 05, DOB 01]. Useful thermal parameters for agricultural applications, for example, or climate change studies can be mapped at local scale according to previously established models. Then, the calculation considered areas (for example, subjected to a given temperature) is easy.

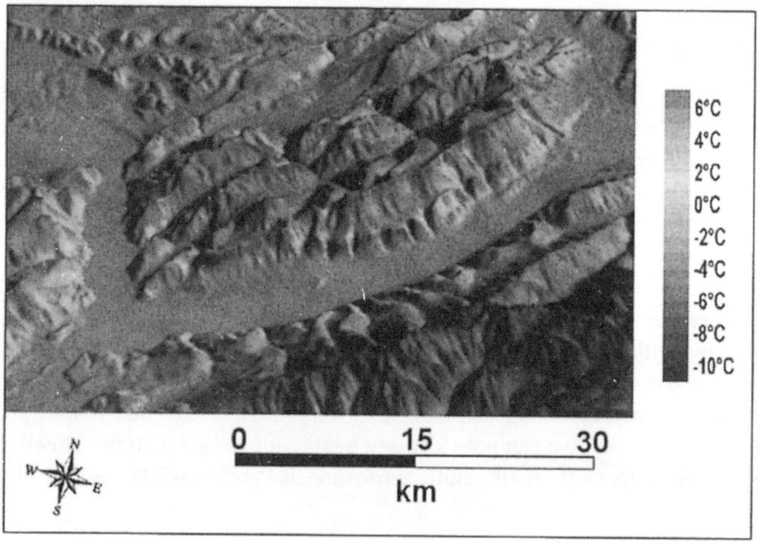

Figure 9.4. *Minimum temperature near Grenoble (4th December 1995)*

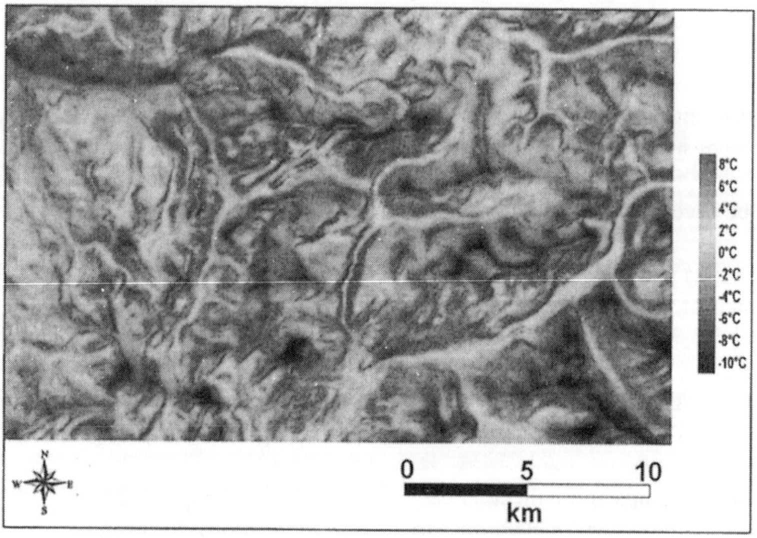

Figure 9.5. *Maximum temperature (3rd December 1995)*

9.5. Conclusion

The main interest in mapping temperature fields, even if an inevitable margin of error still exists, can be found in a more obvious vision of thermal events, in particular at local scale. Many events are indeed not often well readable on maps with point representation. However, interpolating air temperature on a daily scale and in a mountainous area is not easy. Residual kriging provides the best results, but it is not always possible to obtain an exact estimate over a wide area.

We focused on the winter period, which is the most difficult to model (temperature inversion, stratus in valleys, etc.). The results of the stepwise regression can be improved by using low-pass filters on predictor grids and local topographic parameters. The correlation with weather types must also be integrated in a model for a precise interpolation. Other techniques can be tried, like combining different models according to elevation (one equation under temperature inversion, another above although the elevation of the inversion is not always the same over the whole studied area). The solution probably consists of adding GIS methodologies and techniques with high resolution physical or climatologic models. These models are in development, for example, at Meteo France (AROME).

9.6. Bibliography

[BÉN 95] BÉNICHOU P., 1995, *Classification automatique de configurations météorologiques sur l'Europe occidentale*, Monographie no. 8, Météo-France, service central de la communication et de la commercialisation.

[BÉN 87] BÉNICHOU P., LE BRETON O., 1987, "Prise en compte de la topographie pour la cartographie des champs pluviométriques statistiques", *La Météorologie* 7th series, 1987.

[CAR 94] CARREGA P., 1994, Topoclimatologie et habitat. *Analyse spatiale quantitative et appliquée*. Revue de Géographie du laboratoire d'analyse spatiale Raoul Blanchard, UFR Espace et Culture, Nice Sofia Antipolis University, no. 35-36.

[CHA 03] CHAPMAN L., THORNES J.E., 2003, "The use of geographical information system in climatology and meteorology", *Progress in Physical Geography*, 27.3, London, Arnold Publishers, pp. 313-330.

[DOB 01] DOBESCH H., TVEITO O.E., BESSEMOULIN P., 2001, *Geographical Information Systems in Climatological Application,* KLIMA Report 13/01, EUMETNET ECSN, Norwegian Meteorological Institute, Oslo.

[ESC 81] ESCOURROU G., 1981, *Climat et environnement, les facteurs locaux du climat*, Masson, Paris.

[GIE 01] GROUPE D'EXPERTS INTERGOUVERNEMENTAL SUR L'EVOLUTION DU CLIMAT, 2001, *Bilan 2001 des changements climatiques: Les éléments scientifiques*. Rapport du Groupe de travail I du GIEC. OMM PNUE.

[LHO 03] LHOTELLIER R., 2003, *Influence of Geographic Factors on Winter Minimal Temperatures in the French Alps*, International GIS2003 Symposium on GIS and Remote Sensing: Environmental Applications, ed.: Nicolas R. Dalezios, European Commission/ University of Thessaly, Volos, Greece.

[LHO 04] LHOTELLIER R., 2004, *Spatial interpolation of daily minimum and maximum temperatures in the French Alps*, Proceedings of the Conference on the Spatial Interpolation Techniques in Climatology and Meteorology, HMS.

[LHO 05a] LHOTELLIER R., 2005. *Spatialization des températures en zone de montagne alpine*, PhD Thesis, Joseph Fourier University, Grenoble.

[LHO 05b] LHOTELLIER R., PATRICHE C.-V., 2005, *Short-Term Scientific Mission COST 719: France to Romania.*

[SAI 70] SAINTIGNON (de) M.-F., DOUGUEDROIT A., 1970, "Méthode d'étude de la décroissance des températures en montagne de latitude moyenne: exemple des Alpes Françaises du Sud", *Revue de géographie alpine*, vol. LVIII-3, pp. 453-472.

[SAI 76] SAINTIGNON (de) M-F., 1976, "Décroissance des températures en montagne de latitude moyenne: exemple des alpes françaises du nord", *Revue de géographie alpine*, vol. LXIV, pp. 483-494.

[TVE 00] TVEITO O.E. *et al.*, 2000, *Nordic temperature maps*, KLIMA Report 09/00, NORDKLIM, Nordic co-operation within climate activities, Norwegian Meteorological Institute, Oslo.

[TVE 02] TVEITO O.E., SCHONER W., 2002, *Applications of spatial interpolation of climatological and meteorological elements by the use of geographical information systems (GIS)*, KLIMA Report 28/02, NORDKLIM, Nordic co-operation within climate activities, Norwegian Meteorological Institute, Oslo.

[TVE 03] TVEITO O.E., USTRNUL Z., 2003, *A review of the use of large-scale atmospheric circulation classification in spatial climatology*, KLIMA Report 10/03, Det Norske Meteorologisk Institutt, Oslo.

[UST 05] USTRNUL Z., CZEKIERDA D., 2005, "Application of GIS for the development of climatological air temperature maps: an example from Poland", *Meteorological applications*, vol. 12, no. 1, Royal Meteorological Society, pp. 43-50.

[VAN 01] VAN DAM O., 2001 *Forest filled with gaps. Effects of gap size on water and nutrient cycling in tropical rain forest. A study in Guyana.* PhD Thesis Utrecht University. Tropenbos-Guyana Series 10, Tropenbos-Guyana Programme, Georgetown, Guyana.

Chapter 10

Aspects Concerning the Spatialization of Radiation Balance Components

10.1. Introduction

This chapter discusses different conceptual models which can be used for the spatialization of the radiation balance components within GIS. We focus on the models of deriving the global radiation on horizontal surfaces (the Angström formula, models based on the parameterization of clear and cloudy skies radiation) and on the methods of accounting for slope and aspect influence on this parameter (sinus ratio using mean daily sun angle, Kaempfert-Morgen nomogram, SAGA GIS correction). We also discuss some of the possibilities of deriving the land surface albedo from satellite images, which can be further used to estimate the net incoming shortwave radiation and the Stefan-Boltzmann law used for the spatialization of the net outgoing longwave radiation. Examples of applications are presented for a plateau region situated in eastern Romania.

Generally, the number of stations recording radiation fluxes is too small to attempt a spatialization based on statistical interpolation methods. Such spatializations are usually performed by applying a certain conceptual model. Several such models can be used to estimate the *global radiation* on a horizontal surface (the Angström formula which is used, for example, within FAO crop growth models to estimate the potential evapotranspiration; the models based on the parameterization of clear and cloudy skies radiation, which is included in various

---

Chapter written by Cristian Valeriu PATRICHE.

GIS software[1]) and several methods can be further applied in order to account for the spatial differences induced by slope angles and exposition of land surface (sinus ratio using mean daily sun angle, Kaempfert-Morgen nomogram, SAGA GIS correction). Also satellite imagery combined with ground measurements provides good spatial models of global radiation [STR 01].

The Angström formula estimates the global radiation by reducing the radiation amount received at the upper limit of the atmosphere according to a factor depending on the relative sunshine duration and some coefficients reflecting the ratios of direct and diffuse radiations:

$$R_g = R_a [a + b(n/N)] \qquad [10.1]$$

where:

– R_g: global radiation;

– R_a: extraterrestrial radiation, received at the upper limit of the atmosphere, on a surface parallel with the horizon of the determination point;

– n/N: relative sunshine duration, representing the ratio of the actual duration of sunshine (n), to the maximum possible duration of sunshine (N);

– a, b: coefficients expressing the qualitative structure of the global radiation (direct and diffuse radiation).

Subsequently, *the net incoming shortwave radiation* (R_{ns}) can be derived by integrating the albedo (a) into the above formula:

$$R_{ns} = (1\text{-}a)\, R_g \qquad [10.2]$$

Where radiation data is available, the a and b coefficients can be determined by linear regression. Otherwise we have to use predetermined values. For temperate regions the following values are recommended: a = 0.25, b = 0.5.

---

1 Such as *SolarAnalyst*, an ArcView GIS extension module [FU 00]; *solar radiation* module form SAGA (System for Automated Geoscientific Analyses) GIS, developed by *Conrad*; *r.sun* module from PVGIS (PhotoVoltaic Geographical Information System), an open source GIS developed by *Dunlop, Šúri, Huld, Cebecauer*, which is based on the database from the European Solar Radiation Atlas (ESRA) [SCH 00].

The extraterrestrial radiation (R_a) values for different latitudes can be extracted from special tables or they can be calculated by using the formula:

$$R_a = \frac{24(60)}{\pi} G_{sc} d_r [\omega_s \sin(\varphi)\sin(\delta) + \cos(\varphi)\cos(\delta)\sin(\omega_s)] \quad [10.3]$$

where:

– G_{sc}: solar constant (0.0820 MJ/m^2 min or approx. 1.98 kcal/cm^2 min);

– d_r: inverse relative distance Earth-sun;

– ω_s: sunset hour angle (rad);

– φ: latitude (rad);

– δ: solar declination (rad).

The maximum possible sunshine duration (N) can also be extracted either from special tables, for different latitudes, or calculated as follows:

$$N = \frac{24}{\pi} \omega_s \quad [10.4]$$

where ω_s is the sunset hour angle in radians.

The Angström formula does not account for the influence of exposition and slopes, which is very important. One way to integrate these effects is to apply a *sinus ratio correction* in the following manner [PAT 05]:

$$R_g = R_a \left[a + b \frac{n}{N} \frac{\sin(h_m \pm \alpha)}{\sin(h_m)} \right] \quad [10.5]$$

where:

– α: slope angle of land surface;

– h_m: mean daily sun angle estimated by using the latitude (φ) and the number of days in the year (J) according to the formula [ALL 98]:

$$sin(h_m) = sin[0.85 + 0.3\ \varphi\ sin(2\pi\ J/365 - 1.39) - 0.42\ \varphi^2] \quad [10.6]$$

We notice that the effect of the slope and orientation was integrated so as to affect only the direct component of the global radiation. The sign of slope angle is + for southern orientations and – for northern orientations. Therefore, the above formula is only valid for the particular situations of southern and northern orientations. We assume that for the intermediate orientations the incoming shortwave radiation also has intermediate values. Consequently, we have to transform the orientation values, initially ranging from 0° to 360°, into values ranging from 0° (N) to 180° (S). Next, the transformed orientation values will be divided by 180 in order to obtain weighting coefficients (c) ranging from 0 to 1. The formula used for estimating the global radiation received by real terrain is the following:

$$R_g = R_{g\text{-}N} + c\,(R_{g\text{-}S} - R_{g\text{-}N}) \qquad [10.7]$$

where:

– $R_{g\text{-}S}$, $R_{g\text{-}N}$: the global radiation for southern and northern orientations.

Therefore, in order to estimate the global radiation at regional or local scale (where Ra and a, b coefficients can be considered constant), we need a DEM for deriving slopes and aspects and a spatialization of n/N ratio (by regression or residual kriging). An example is shown in Figure 10.1 for a plateau region (930 km^2) in eastern Romania (see the color plates section for Figure 10.1).

The mean annual sunshine hours involved in the calculation of the global radiation were estimated by regression, but the predictor was the first principal component's scores obtained by condensing the altitude-latitude-longitude information. This synthetic predictor was proved to perform better than the three initial predictors taken separately.

Following the steps described above to derive the global radiation may be a bit time consuming. Therefore, we wrote a simple script in TNTmips Spatial Manipulation Language (SML), which requires as input the slopes, the transformed orientations (N-S component) and the actual sunshine hours in order to derive the global radiation.

Another way to account for slope and aspect in the determination of the global radiation at temperate latitudes is to use the *Kaempfert-Morgen nomogram*. By entering the exposition class and the slope of the landform, the chart gives us the annual global radiation. Since the chart starts from a fixed annual value for a horizontal surface, it would be useful to calculate the relative values in order to be able to accommodate different horizontal values (Figure 10.2).

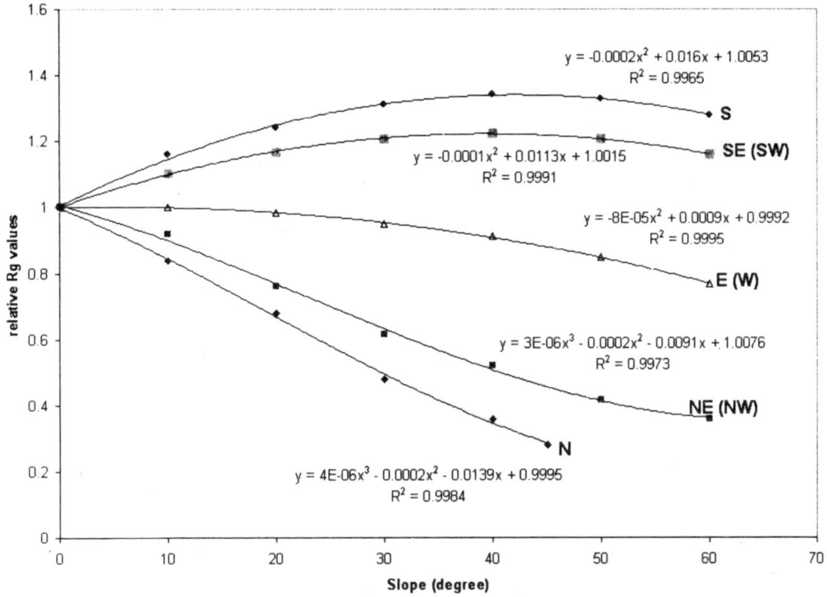

Figure 10.2. *Kaempfert-Morgen nomogram for annual relative values*

The Kaempfert-Morgen nomogram applies a more generalized correction for the horizontal global radiation because it uses exposition classes instead of numerical aspect values and also because it was designed for middle latitudes. In order to overcome the former, we wrote a script by means of which we applied a linear variation of the global radiation values between N-NE (NW), NE (NW)-E (W), E (W)-SE (SW) and SE (SW)-S expositions.

Another way to deal with the spatialization of the global radiation is to use the *parameterizations for clear skies* (I_{clear}) and *cloudy skies* (I_{cloudy}) solar radiation [ENT 97]:

$$I_{clear} = I_0\, e^{-n \alpha_1 m} \qquad [10.8]$$

$$I_{cloudy} = I_{clear}\,(1 - 0.65N^2) \qquad [10.9]$$

where:

– I_0: extraterrestrial radiation, which is received at the upper limit of the atmosphere, on a surface parallel to the horizon of the determination point;

– n: atmospheric turbidity (n = 2 for very clean and clean air, n = 5 for thick smog conditions;

– α_l: molecular scattering coefficient parameterized as $\alpha_l = 0.128 - 0.054 \log_{10}(m)$;

– m: optical air mass (approximately $1/\sin\alpha$, where α is the mean sun angle);

– N: fractional cloud cover.

As in the case of the Angström formula, I_{cloudy} can be corrected in two ways in order to account for slope and aspect influence, either by using the sinus ratio or the KM nomogram for annual values. Since I_{cloudy} is the global radiation reaching the Earth surface, it must be divided into its diffuse and direct components in order to apply the correction only to the direct component.

The global radiation can be more easily assessed by using the solar radiation module of the *SAGA GIS software*, developed by *Conrad* (Figure 10.3). However, it should be mentioned that this module does not work for version 1.2.

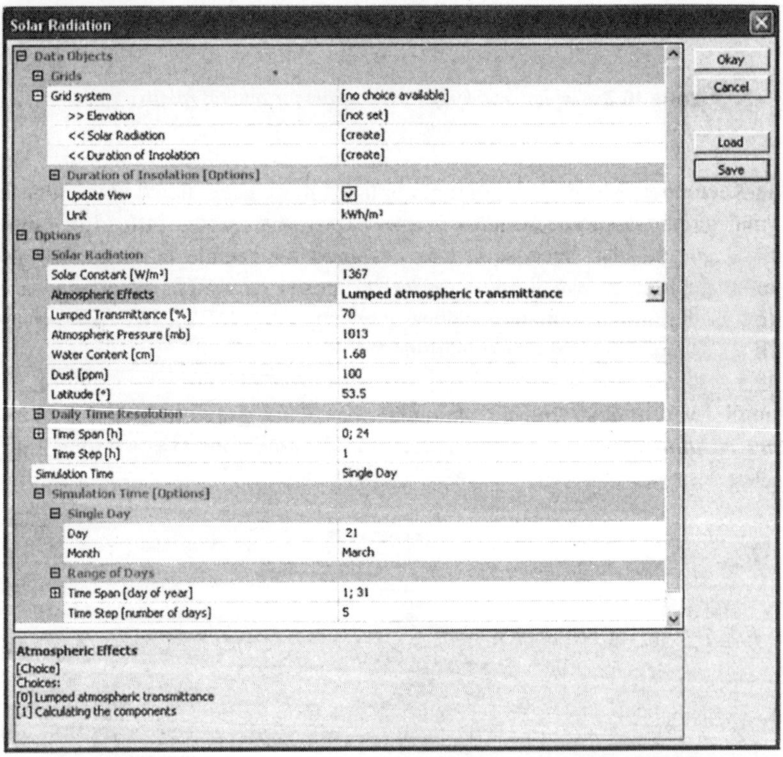

Figure 10.3. *Solar radiation module interface in SAGA GIS software*

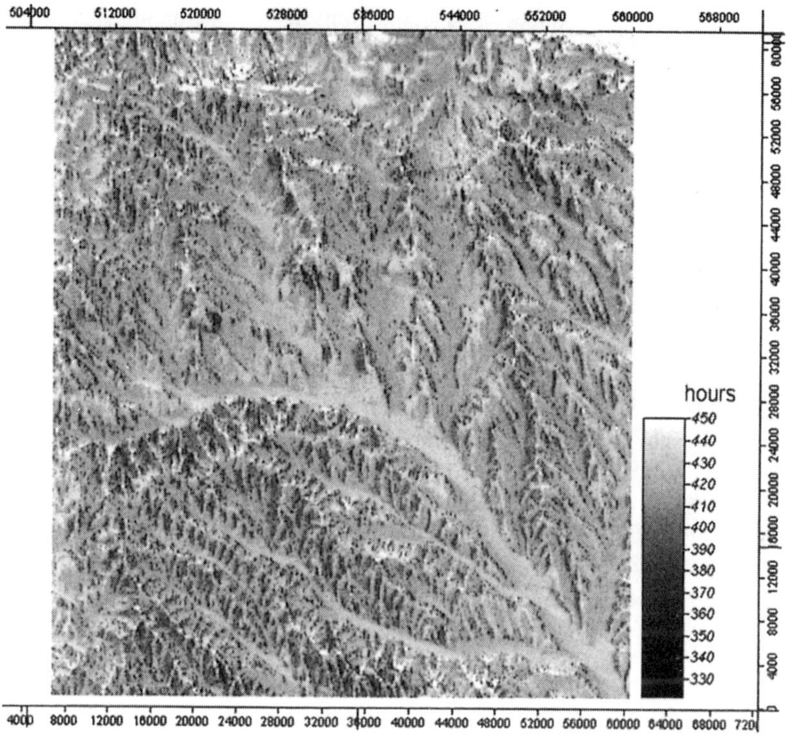

Figure 10.4. *Astronomically possible sunshine duration in July (hours)*

The solar radiation model included in the beta version 2 uses a parameterization for clear skies radiation which requires as input the DEM of the region studied, the lumped atmospheric transmittance, atmospheric pressure, water content, dust content and latitude. Also, the value of the solar constant can be changed. The resulting spatialization can be expressed either in kWh/m^2 or kJ/m^2. The clear skies radiation can be estimated for a single day (or even less, for a few hours), for several days (for example, a month) or for a whole year. Choices can be made on time step to be used (for example, calculate radiation every 2 hours and every 2 days from a certain month). A finer time step will produce a more accurate spatialization but it will also take more time. Once all the parameters are set, the module generates successive spatializations of the clear skies radiation for each time step and adds them all up at the end to produce the final spatialization.

One of the shortcomings of the module is that it does not produce a cloudy skies radiation. However, this can be obtained later, for example, by multiplying the results by $(1\text{-}0.65N^2)$ as in the above formula.

The module also produces a spatialization of the astronomically sunshine duration based on the relief configuration, which is very useful for a more realistic determination of the insolation fraction, especially in fragmented terrain (mountainous areas) where the astronomically sunshine duration cannot be considered constant. An example from eastern Romania is given in Figure 10.4. We may notice higher values on ridges where the horizon is wider and lower values in deep shaded valleys.

10.2. Comparison of the models

For comparison, we estimated the global radiation on a horizontal surface by using the models described above and then we compared the results with the real mean monthly global radiation data recorded at the Iasi meteorological station in Romania. We notice that the Angström formula gives the best estimate of the global radiation (Figure 10.5 and Table 10.1). However, the values are a little underestimated during spring, especially in March, whereas the best estimated month is July. The general underestimation of radiation values, which is the case for all models used, leads to underestimated annual sums, with a difference ranging from 2.57 kcal/year in the case of the Angström formula, to 11.78 kcal/year in the case of I_{cloudy} parameterization.

We may notice that the I_{cloudy} parameterization underestimates the global radiation values in all the months, whereas the SAGA model overestimates the values from April to June. An interesting fact is that the best estimate is found for July, when all models give very similar results.

Even though the use of the Angström formula is the best choice in our example, the situation may be quite different in different climatic conditions, where other models or other Angström coefficients may perform better.

In the case of the methods used for slope and aspect correction, we cannot use any real data because we do not dispose of the measurement of global radiation on different slopes and expositions. Logically, the SAGA ray tracing model is the best because it actually generates the solar incidence angles for successive hours within a day, from which the daily global radiation sums are then calculated. The Kaempfert-Morgen nomogram correction, as we mentioned above, is more generalized. Even if we overcome the disadvantage of using exposition classes instead on numerical exposition values, we still have a degree of generalization due to the fact that it was designed for middle latitudes.

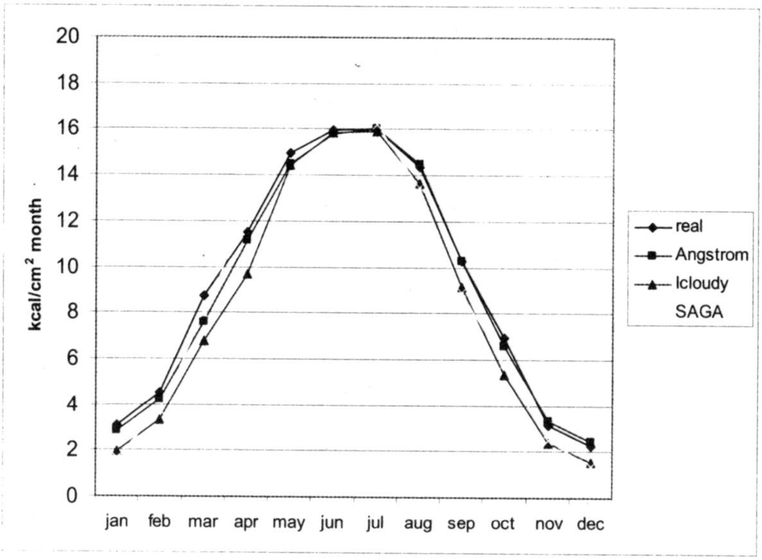

Real and estimated annual regime

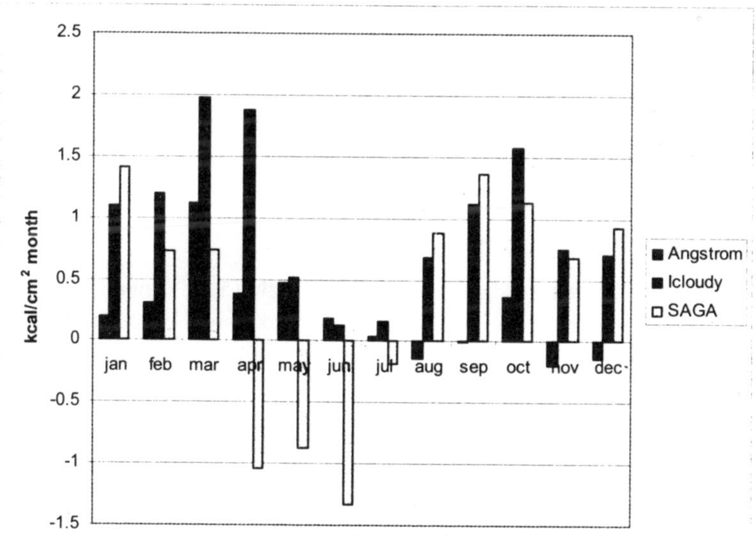

Differences between real and estimated values

Figure 10.5. *Comparison of real and estimated global radiation values on a horizontal surface for the Iasi meteorological station (eastern Romania, 46.17° latitude, 102 m altitude)*

Models	**Maximum error**		**Minimum monthly error**	**Mean annual error**	**Root mean annual square error**
	Highest monthly overestimation	Highest monthly underestimation			
Angström	–0.192	1.115	-0.006	0.215	0.407
I_{cloudy}	–	1.973	0.130	0.981	1.140
SAGA	–1.334	1.403	-0.180	0.368	0.998

Table 10.1. *Error parameters for the global radiation models*

The sinus ratio correction has several shortcomings:

– It does not sum the hourly radiation values.

– It presents Potential problems for steep slopes:

– South-facing surfaces: if the maximum sun angle (h_{max}, at noon) plus the slope is greater than 90 ($h_{max}+\alpha > 90$), the radiation derived from the sinus ratio correction will be greater than the real one because the sinus ratio uses the mean sun angle and $h_m+\alpha < 90$.

– North-facing surfaces: if $h_m<\alpha$, the direct component of the global radiation must be set to zero, but in fact there is a small amount of direct radiation reaching these surfaces which corresponds to the time when the solar angle is between h_m and h_{max}.

Corrections are needed in these cases of steep slopes occurrences.

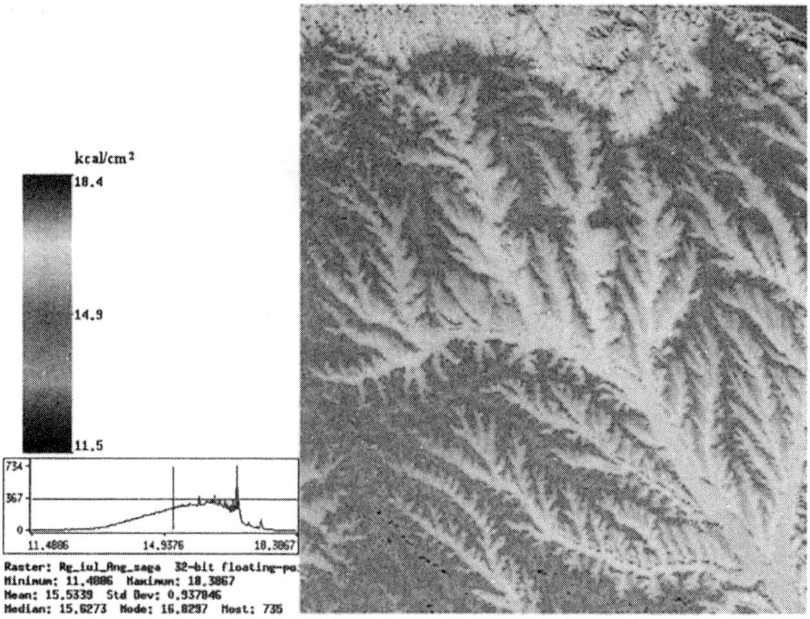

Angström formula – sinus ratio correction

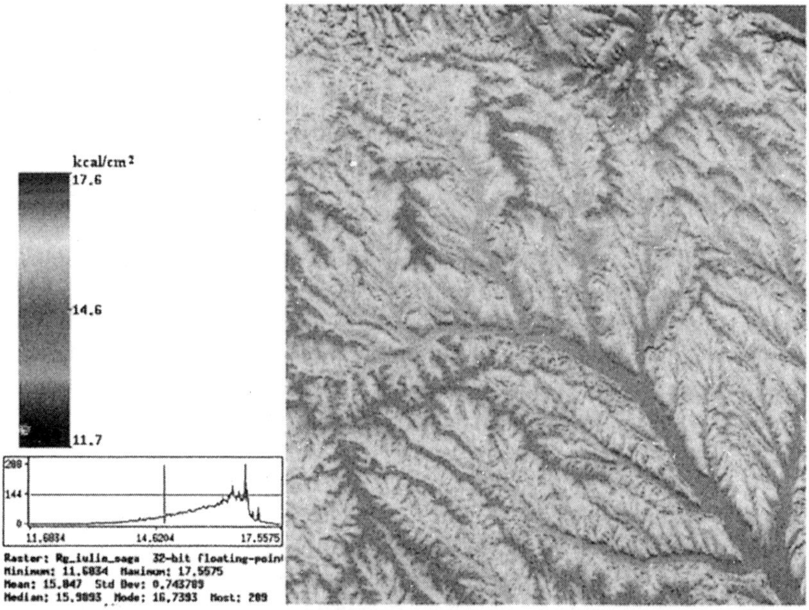

SAGA model

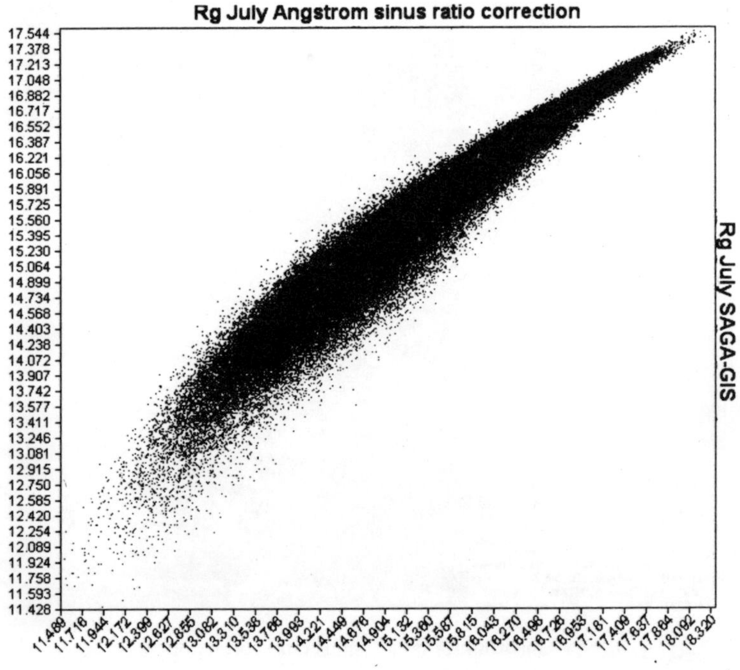

Figure 10.6. *Comparison of global radiation spatialization by using the Angström formula and the SAGA model for July (eastern Romania, 90 m resolution)*

For comparison, we estimated the global radiation for a region situated in eastern Romania for the month of July, when all the models provide similar results on horizontal surfaces (Figure 10.6).

We notice that the values estimated by the Angström formula with sinus ratio correction are generally a little lower than the SAGA estimated values, with the exception of the higher radiation domain, where they become a little higher. The two histograms are quite similar in respect to mean, minimum, mode and median values. However, the shapes and standard deviations indicate a greater dispersion of values in the case of Angström formula with sinus ratio correction than in the SAGA model.

For the annual values, we notice that the Kaempfert-Morgen nomogram correction is more closely correlated with the SAGA model, than with the sinus ratio correction (Figure 10.7). Therefore, an Angström spatialization with KM correction seems to be the best choice for the annual global radiation sums in the case of our region.

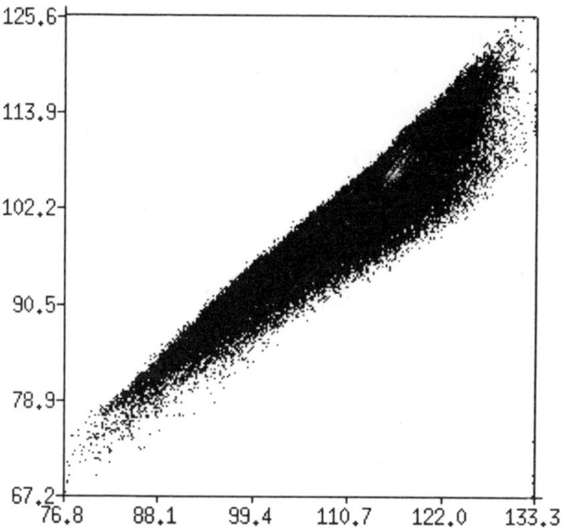

SAGA (X) – Angström with sinus ratio correction (Y)

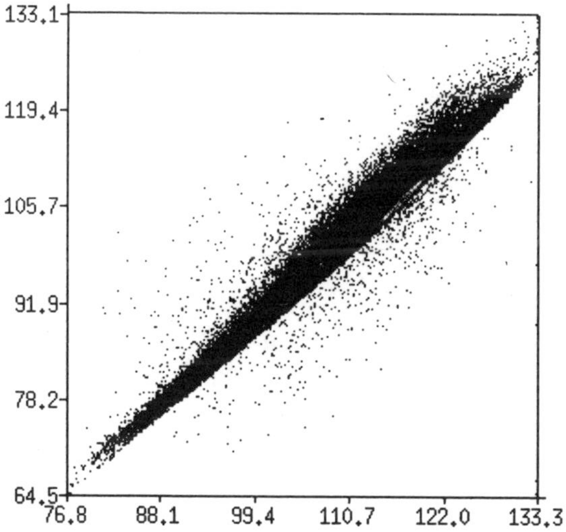

SAGA (X) – Angström with Kaempfert-Morgen correction (Y)

Figure 10.7. *Correlation between annual global radiation spatializations by using different methods of accounting for slope and aspect influence (eastern Romania, 90 m resolution)*

The net incoming shortwave radiation (R_{ns}) is the part of the global radiation which is absorbed by the active surface and subsequently transformed into heat fluxes:

$$R_{ns} = (1 - a)\, R_g \qquad [10.10]$$

Therefore, the annual regime and spatial distribution of R_{ns} depends on all the factors controlling the global radiation plus the spatial and temporal albedo variation.

The spatial distribution of the albedo values can be approached in two ways:

– By attaching *predefined values* to land cover polygons in a land cover map (it is the case of the R_{ns} spatialization in Figure 10.1). This is the easier way, but it implies some degree of generalization due to the use of rigid albedo values, which, in reality, vary within certain limits.

– By deriving the albedo values from *satellite images*. This is more precise but it is also a far more difficult and time consuming task. Several methods of calculation can be used depending on the satellite sensor (Landsat – TM/ETM, NOAA – AHVRR, METEOSAT, etc.). The basic steps for obtaining the albedo values by using Landsat images are [GOI 92]:

- Transformation of the digital numbers (DN) in radiance values by using the sensor's calibration parameters (gain, offset):

$$L = Offset + (Gain\ x\ DN) \qquad [10.11]$$

- Calculation of apparent reflectance (ρ) for each spectral band:

$$\rho = \pi\, L\, d^2 / E \cos(\theta_s) \qquad [10.12]$$

where L is the spectral radiance, d is the Earth-sun distance correction, E is the mean solar exoatmospheric irradiance and θ_s is the sun zenith angle.

- Calculation of albedo for the whole visible (TM1,2,3; ETM1,2,3), near (TM4; ETM4) and middle infrared (TM5,7; ETM5,7) spectrum:

$$a = \sum \rho_i E_i / \sum E_i \qquad [10.13]$$

The net outgoing longwave radiation (R_{nl}) is the difference between the Earth's thermal infrared emission and the Earth oriented atmospheric longwave radiation. It can be estimated according to the *Stefan-Boltzmann* law, on the basis of actual water vapor pressure and relative sunshine duration:

$$R_{nl} = \sigma T^4 \left(0.56 - 0.079\sqrt{e_{act}}\right) \cdot \left(0.1 + 0.9\, n/N\right) \qquad [10.14]$$

where:

– σ: *Stefan-Boltzmann* constant (4,903 · 10^{-9} MJ/m^2 K^4 day);

– T: absolute temperature (K);

– e_{act}: actual water vapor pressure (mb);

– n/N: relative sunshine duration.

If the data concerning the actual water vapor pressure is not available, we can estimate this parameter by using air temperature and relative humidity, according to the formula [ALL 98]:

$$e_{act} = e_{sat}(r/100) = 6.1078 e^{[17.27t/(t+237.3)]}(r/100) \qquad [10.15]$$

where:

– e_{sat}: saturation vapor pressure (mb);

– r: relative humidity (%);

– t: air temperature (°C).

According to the formula above, the annual regime and the spatial distribution of the net outgoing longwave radiation are mainly determined by air temperature, relative humidity and relative sunshine duration.

The net radiation (radiation balance) (R_n) is the difference between the net incoming shortwave radiation (R_{ns}) and the net outgoing longwave radiation (R_{nl}):

$$R_n = R_{ns} - R_n \qquad [10.16]$$

The positive values of the net radiation indicate an energy gain of the land surface which dissipates under different forms, which contribute to the warming of the soil, down to a certain level, to the warming and the movement of the air above and to water evaporation. The negative values indicate an energy deficit, which is compensated by the climatic system by using the energy surplus from the surrounding areas with positive radiation balance.

10.3. Bibliography

[ALL 98] ALLEN R.G., PEREIRA L.S., RAES D., SMITH M., *Crop evapotranspiration – Guidelines for computing crop water requirements*, FAO Irrigation and drainage paper, no. 56, Rome, 1998.

[ENT 97] ENTEKHABI D., "Land surface processes: basic tools and concepts", Hydrometeorology and Climatology (Marani M., Rigon R., eds.), Environmental Dynamics Series, V, Venice, 1997, p. 3-46.

[FU 00] FU P., RICH P. M., *The Solar Analyst 1.0 User Manual. Draft*, Helios Environmental Modeling Institute, LLC, 2000, http://www.fs.fed.us/informs/download.php.

[GOI 92] GOITA K., ROYER A., "Land surface climatology and land cover change monitoring since 1973 over a north-Sahelian zone using Landsat data", Geocarto-International, no. 2, 1992, p. 15-28.

[OLA 04] OLAYA V., *A gentle introduction to SAGA GIS*, Edition 1.1, Rev. December 9, 2004, http://www.saga-gis.uni-goettingen.de/html/index.php.

[PAT 05] PATRICHE C.V., *The Central Moldavian Plateau Between Vaslui and Stavnic Rivers – A Physical Geography Study*, Terra Nostra Publishing, Iasi, 2005.

[SAH 04] SAHA S. K., "Retrieval of agrometeorological parameters using satellite remote sensing data", Satellite remote sensing and GIS applications in agricultural meteorology: proceedings of the training workshop, 7-11 July, 2003, Dehra Dun, India, WMO, TD no. 1182, 2004, p. 175-194.

[SCH 00] SCHARMER K., GREIF J. (eds.), *The European Solar Radiation Atlas. Vol. 2: Database and Exploitation Software*, Presses de l'École des Mines, Paris, http://www.ensmp.fr/Fr/Services/PressesENSMP/Collections/ScTerEnv/Livres/atlas_tome2.htm.

[STR 01] STRUZIK P., Spatialization of Solar Radiation – draft report on possibilities and limitations, COST action 718: Meteorological applications for agriculture, Budapest, 2001, http://agromet-cost.bo.ibimet.cnr.it/fileadmin/cost718/repository/Buda_Solar.pdf.

Part 3

Demo Projects

Chapter 11

The Use of GIS Applications in Meteorology and Climatology: A Need for the Application of Regional Ecological Modeling Approaches

11.1. Introduction

Geographic Information Systems (GIS) are widely used in geosciences, but in some cases it is still underused in the fields of meteorology and climatology due to the fact that it is actually difficult to integrate highly variable temporal and spatial datasets, such as spatial temperature and rainfall distributions, in traditional GIS systems.

To enhance the use of GIS in climatology and meteorology, the COST719 action "The use of GIS in climatology and meteorology" with three working groups was established in 2001 [DYR 05a]. This chapter tries to give a short summary of the results of working group (WG) 3 within COST719. The main goals of WG3 were:

– Inventory as well as classification of existing GIS applications as a definition of the state of art.

– Identification and exchange of existing useful applications in the fields of meteorology and climate research and some new case studies/prototypes/training packages to be shared between the participants of the action.

Chapter written by Martin WEGEHENKEL.

– Overview of user needs and requirements of such GIS applications from the point of view of spatial modelers like ecologists or hydrologists.

Other aspects like software recommendation and data formats were discussed in WG1 [WEL 05]. Spatial interpolation methods were analyzed and discussed in WG2 [TVE 05].

11.2. Overview of the actual state of the art of GIS applications in meteorology and climatology

This chapter focuses at first on a classification of existing GIS applications and then on new GIS training and education programs. Such a classification and overview was published by [CHA 03] within the framework of WG3. In this overview, a large amount of GIS applications were referenced and summarized. In order to illustrate the findings of WG3 with regard to the contents of this overview concerning the actual state of the art in GIS applications in meteorology and climatology, four demo projects A, B, C and D developed by members of WG3 were established. The applications that have been developed within these Demo projects focus on precipitation, temperature, energy balance and GIS education.

The first demonstration project, DEMO A [DYR 05b], deals with precipitation mapping by using combined information from satellite, mesoscale forecast and synoptic models as well as climatologic ground measurements. The satellite data together with standard synoptic and climatic measurements as well as radar and NWP analyses are merged together for precipitation analysis. The system makes it possible to display the rain field which is forecasted by the numerical weather prediction model and the precipitation which is observed with other ancillary information through a unique front-end software interface [DYR 05b].

The second project, DEMO B, consists of an easy-to-use application for the spatial interpolation of precipitation and temperature, which was developed by Potzmann, ZAMG Wien as an Arcview3.x extension. Actually, the data can be interpolated by the following methods:

– Internal ArcView Procedure Inverse Distance weighted (IDW).

– Internal ArcView Procedure – Spline.

– External procedure – Kriging realized in Fortran program.

– A multi-layer IDW method written in Avenue.

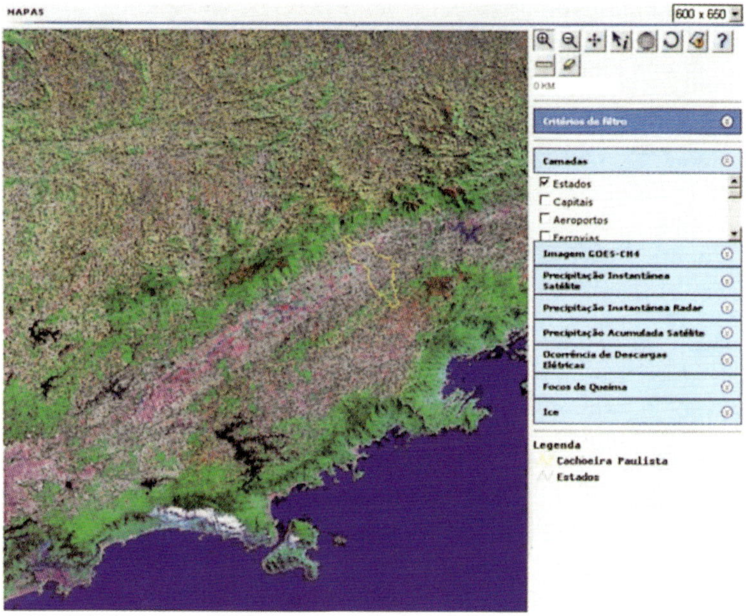

Figure 2.4. *SOS interface for web browser*

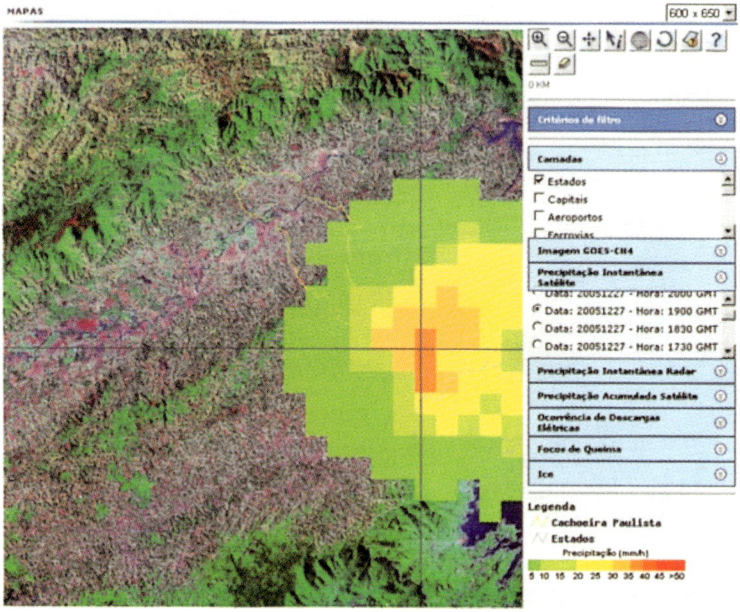

Figure 2.5. *Storm predicted by nowcasting techniques and emission of an alert by the SOS*

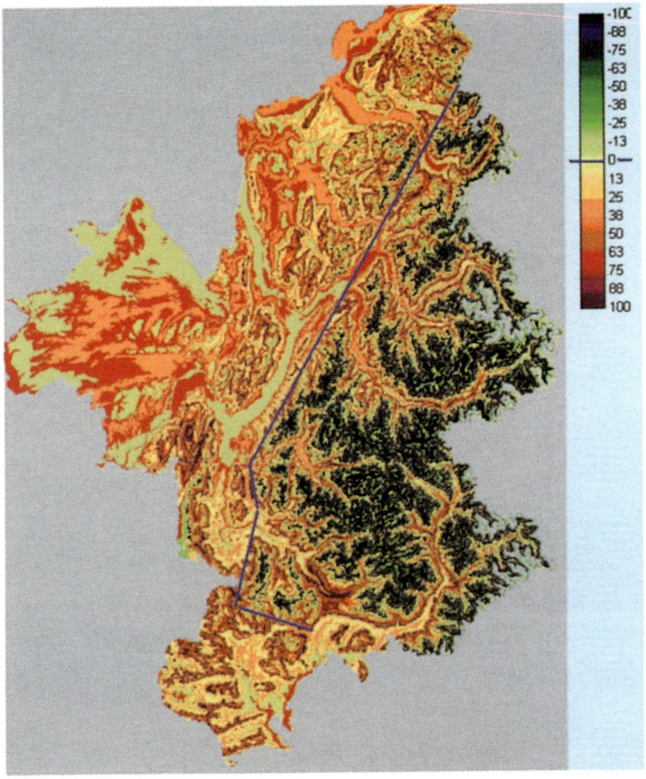

Figure 5.3. *Relative 3D representativeness of the sample at local scale*

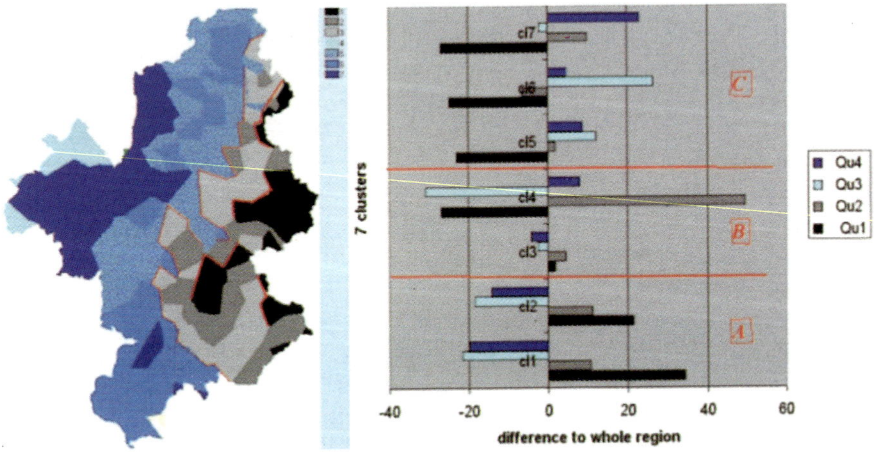

Figure 5.7. *Map and profiles of the 7 groups of Voronoi polygons*

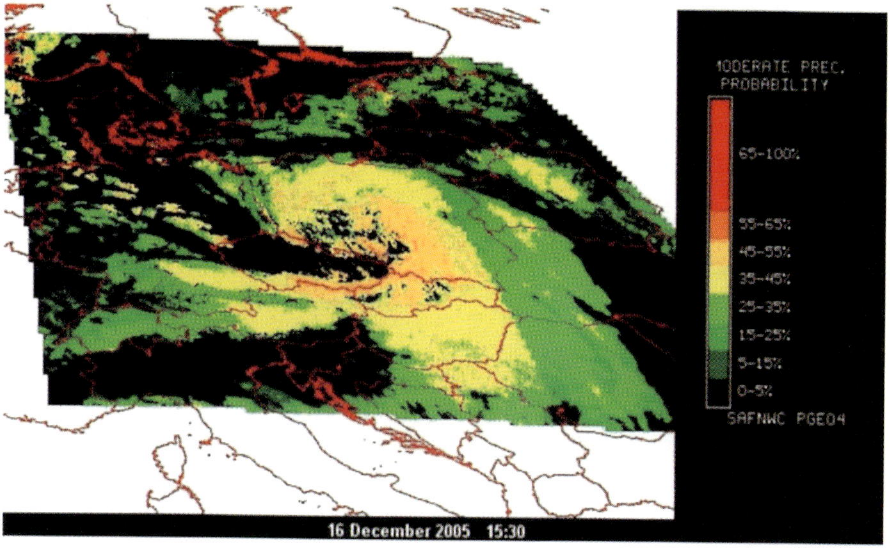

Figure 7.3. *Probability of weak and moderate precipitation on the 16.12.2006 at 15.30 UTC in Poland. "Satellite" projection*

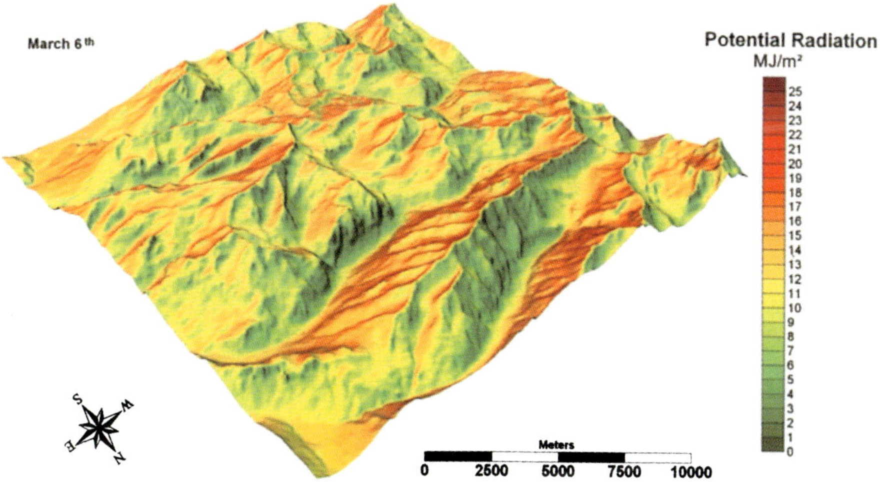

Figure 9.1. *Solar potential radiation estimated by GIS on 6th March*

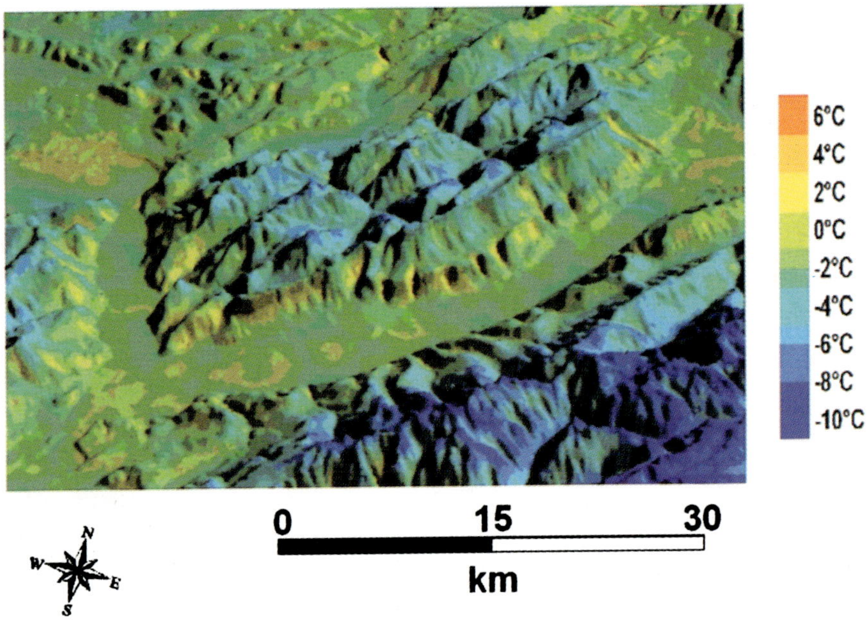

Figure 9.4. *Minimum temperature near Grenoble (4th December 1995)*

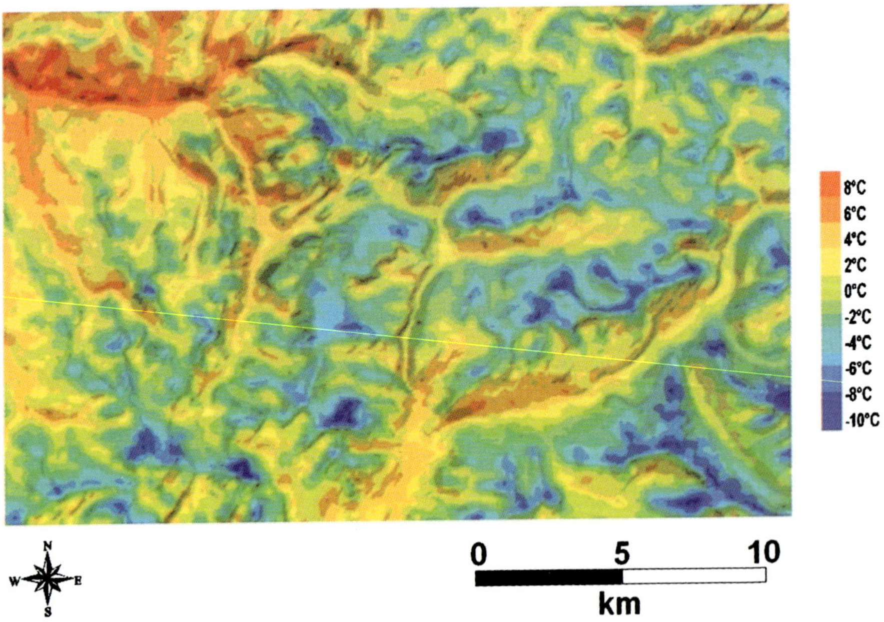

Figure 9.5. *Maximum temperature (3rd December 1995)*

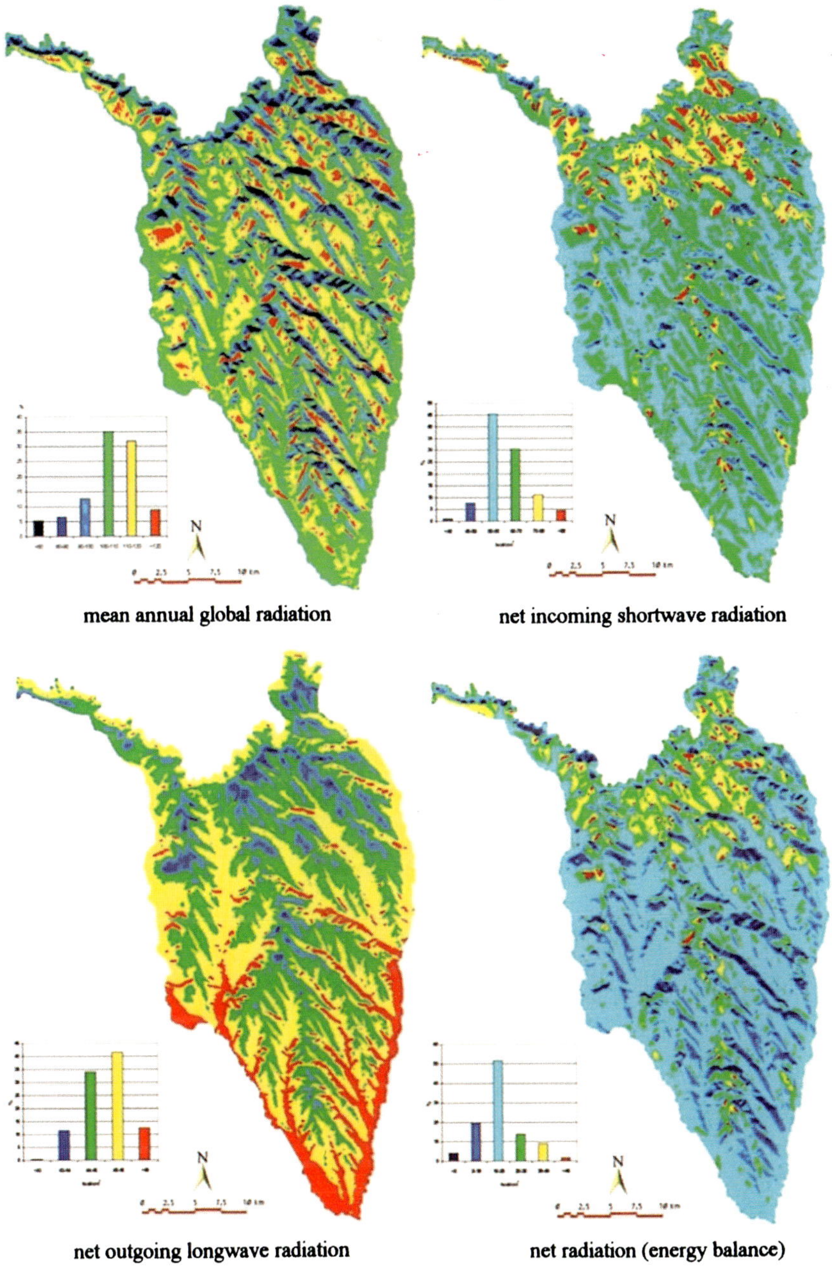

Figure 10.1. *The Central Moldavian Plateau: estimated mean annual values of radiation balance (kcal/cm², resolution 20m) [PAT 05]*

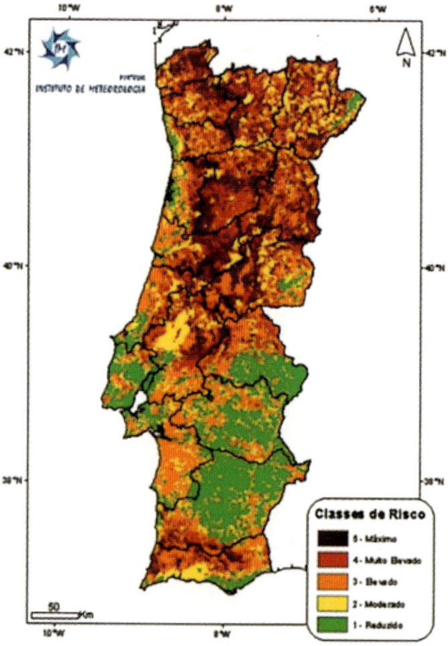

Figure 12.2. *CFI for Portugal mainland in 2006 [ISA 06]*

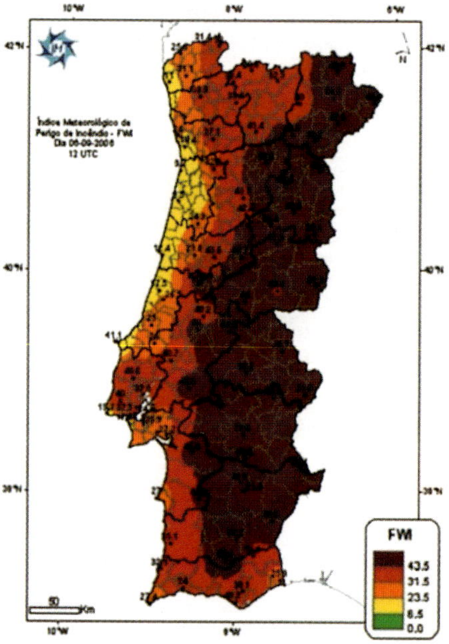

Figure 12.3. *FWI for Portugal mainland, 6th September 2006*
(the labelled points represent the FWI values at the meteorological stations)

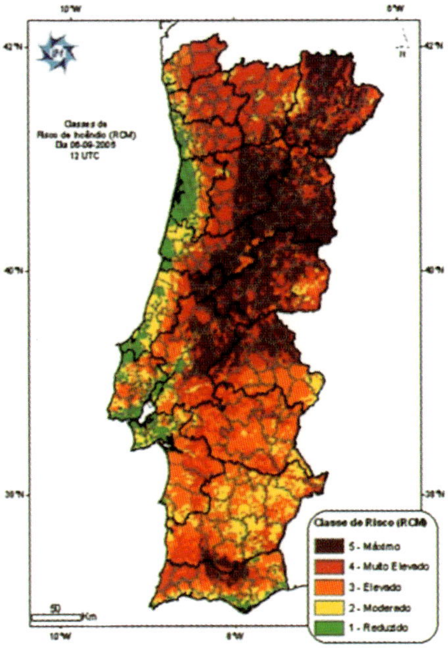

Figure 12.4. *CFR for Portugal mainland, 6th September 2006*

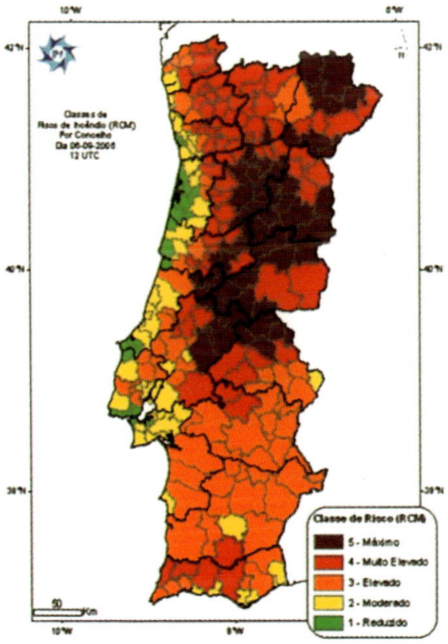

Figure 12.5. *CFR for the municipality level in Portugal mainland, 6th September 2006*

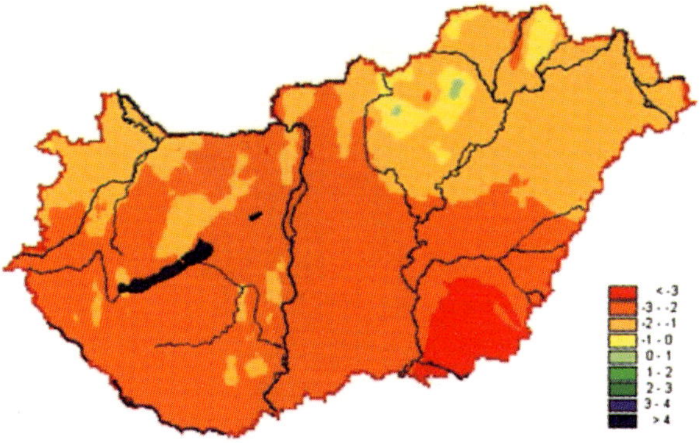

Figure 14.2. *6-month SPI index for Hungary in August 2003*

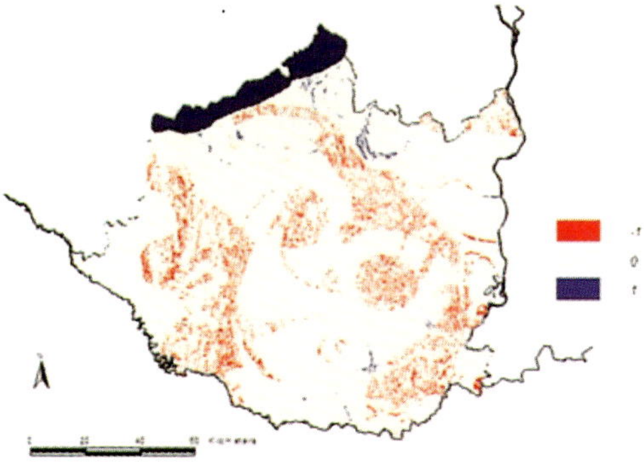

Figure 14.4. *Changes of drought vulnerability according to the distinction of maps of 1951-1980 and 1961-1990*

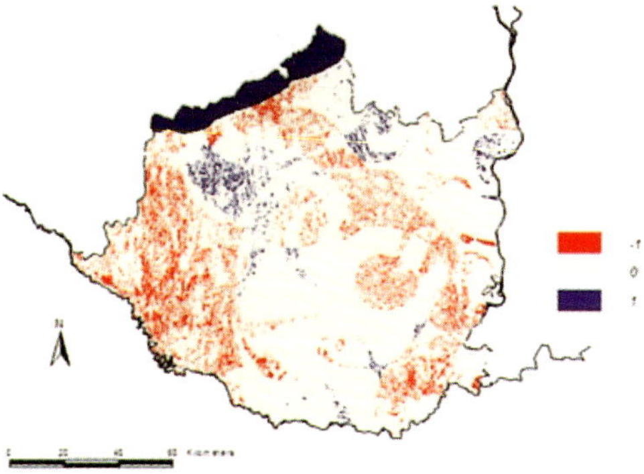

Figure 14.5. *Changes of drought vulnerability according to the distinction of maps of 1951-1980 and 1971-2000*

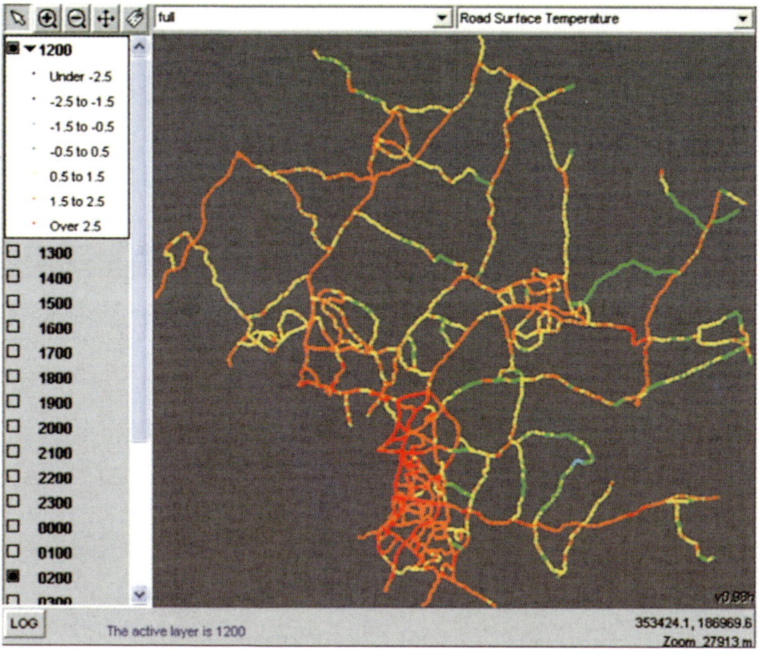
full
Road Surface Temperature
1200
Under -2.5
-2.5 to -1.5
-1.5 to -0.5
-0.5 to 0.5
0.5 to 1.5
1.5 to 2.5
Over 2.5
1300
1400
1500
1600
1700
1800
1900
2000
2100
2200
2300
0000
0100
0200
LOG
The active layer is 1200
353424.1, 186969.6
Zoom 27913 m

a)

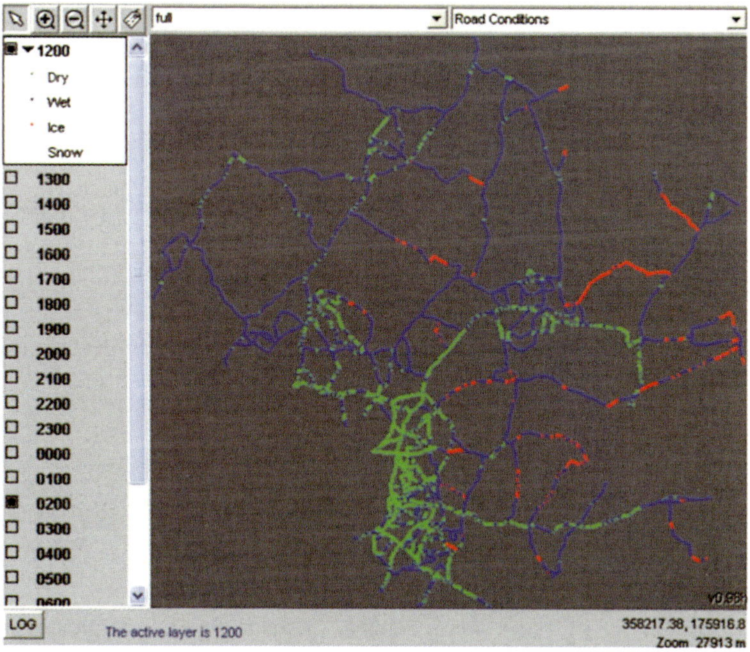
full
Road Conditions
1200
Dry
Wet
Ice
Snow
1300
1400
1500
1600
1700
1800
1900
2000
2100
2200
2300
0000
0100
0200
0300
0400
0500
LOG
The active layer is 1200
358217.38, 175916.8
Zoom 27913 m

b)

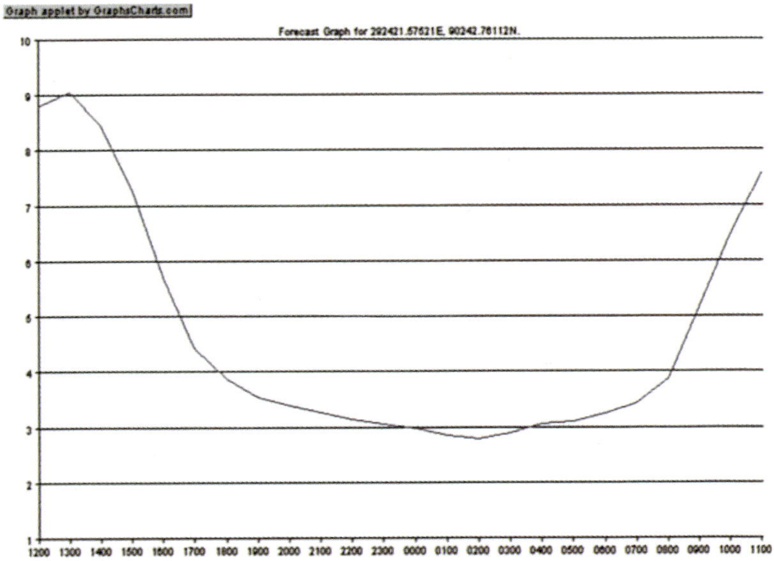

c)

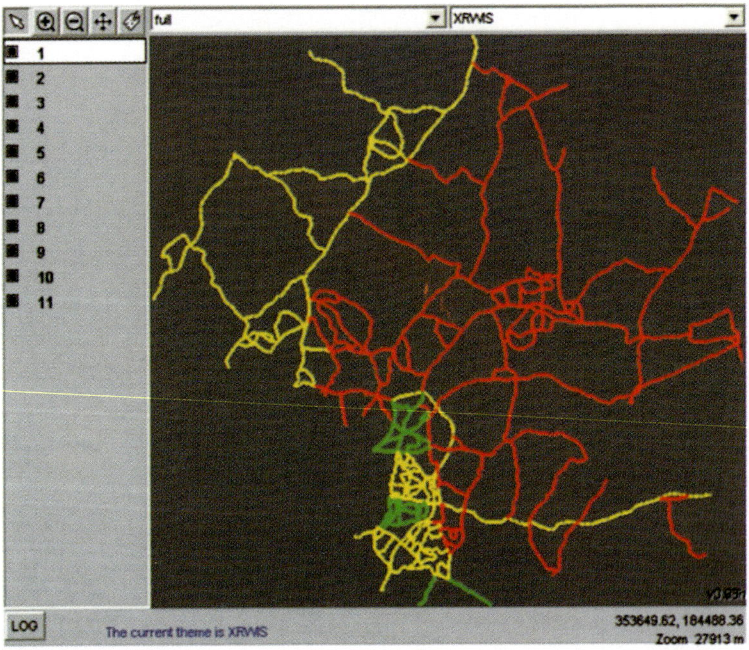

d)

Figure 16.2. *Conceptual visualisation of the new XRWIS paradigm showing: a) road surface temperatures, b) road surface conditions, c) forecast graph and d) "traffic light" salting routes*

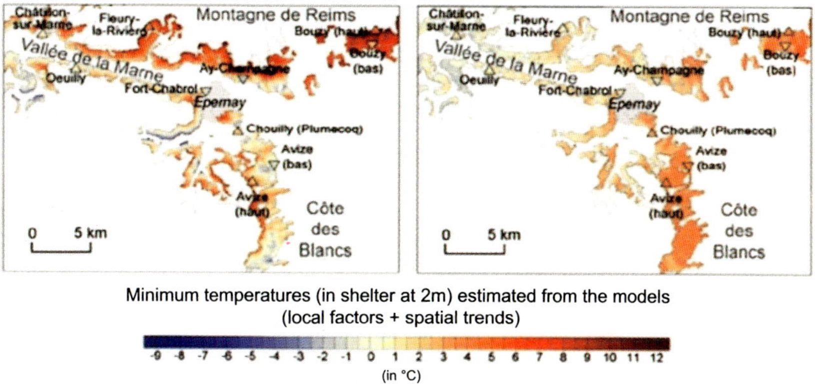

Figure 17.1. *Spatialization of frost hazard in the Champagne vineyard (from [MAD 04]): minimal (night) temperatures estimated from a regression model by using observations, topography from a DEM and land cover*

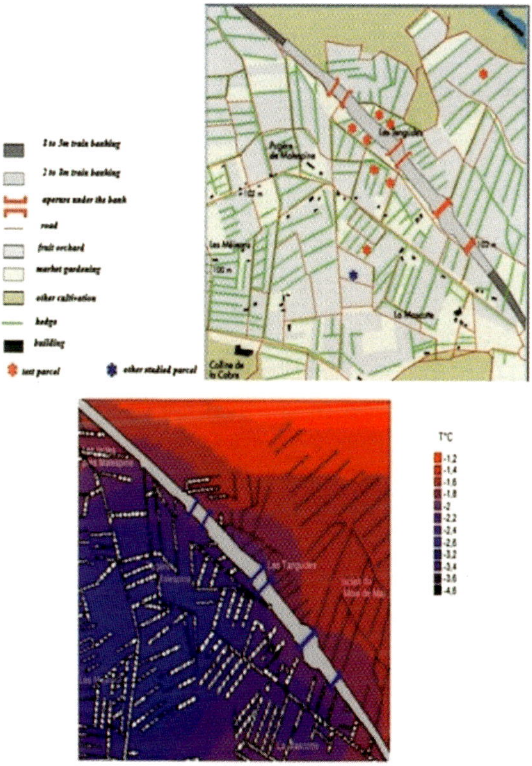

Figure 17.2. *Attempt at parcel-scale spatialization of minimal temperatures along the embankment of high speed train in the Durance valley orchard (from [QUE 02])*

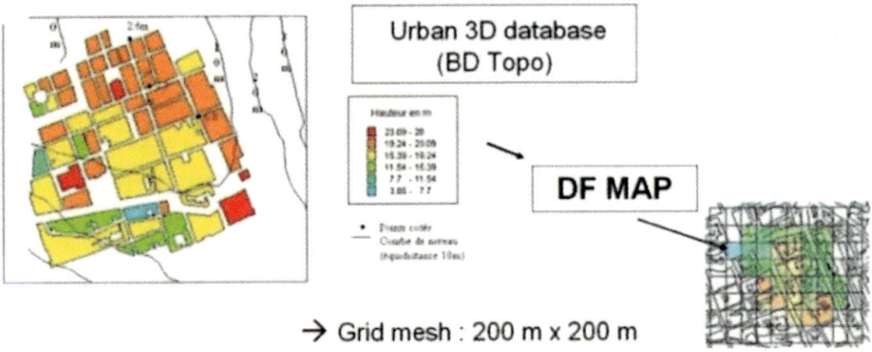

a) From 3D vectorial data to urban parameters maps

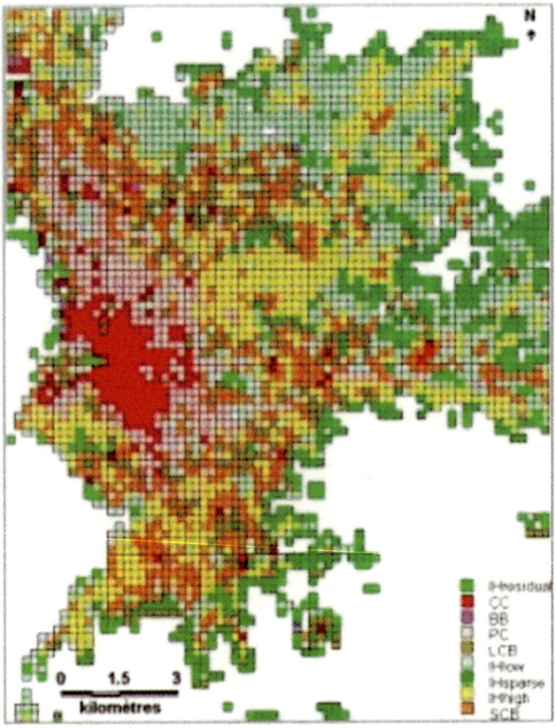

b) *Classification of urban fabric types in Marseilles*

Figure 17.3. *GIS-based analysis of the urban fabric for use in urban climate studies (from [LON 03])*

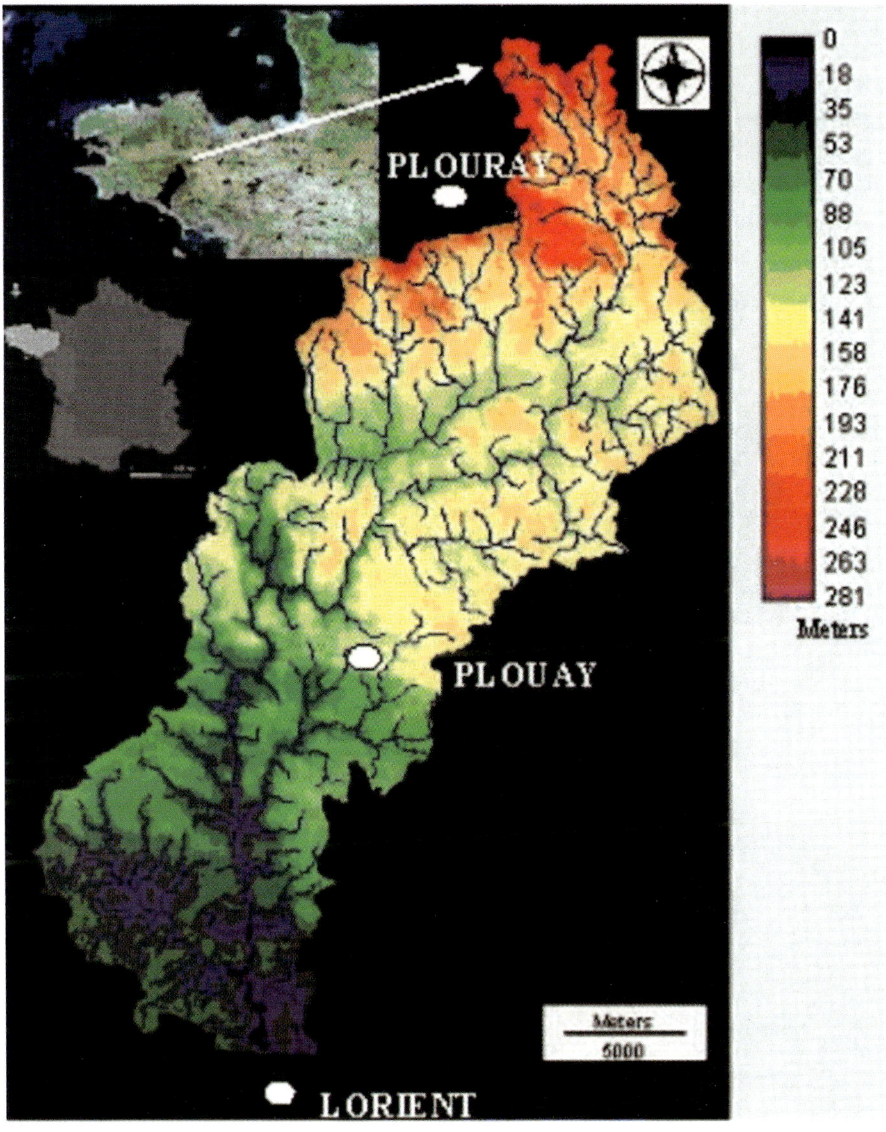

Figure 18.1. *Localization of the area of study*

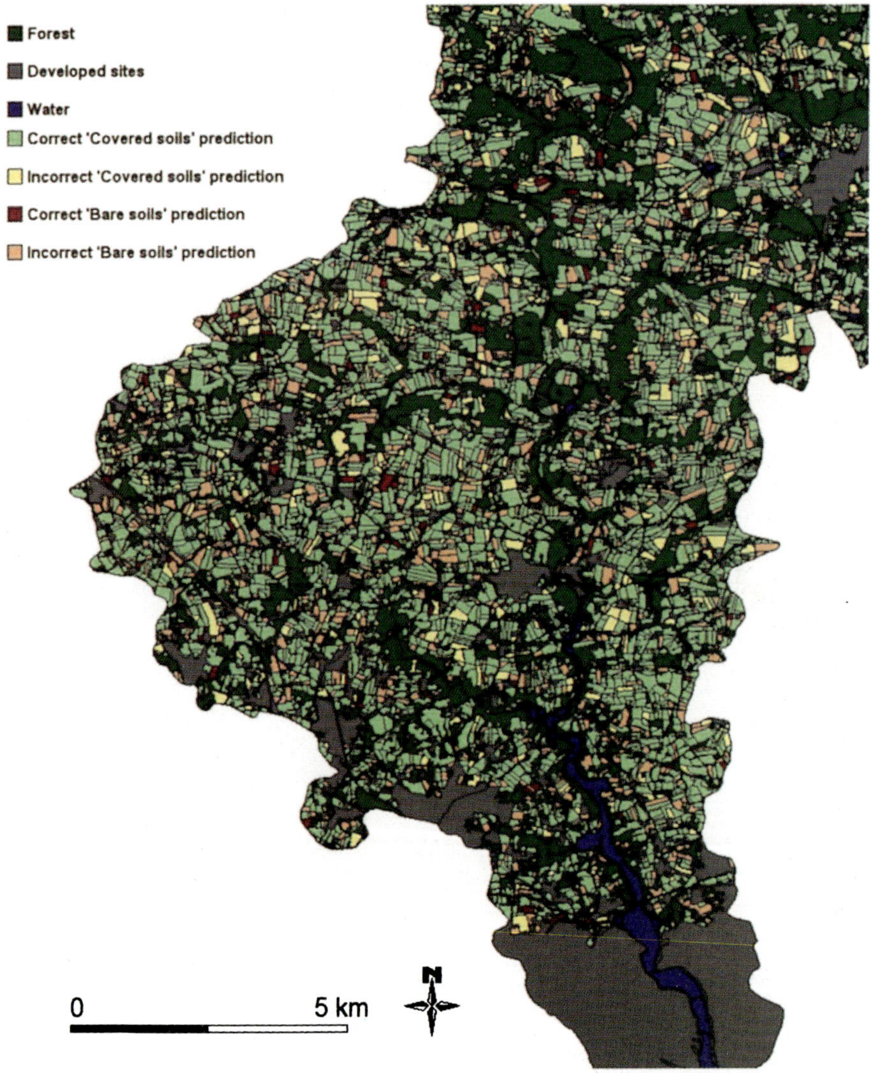

Figure 18.5. *Land cover prediction for the winter of 2003-2004 (south of the Scorff watershed)*

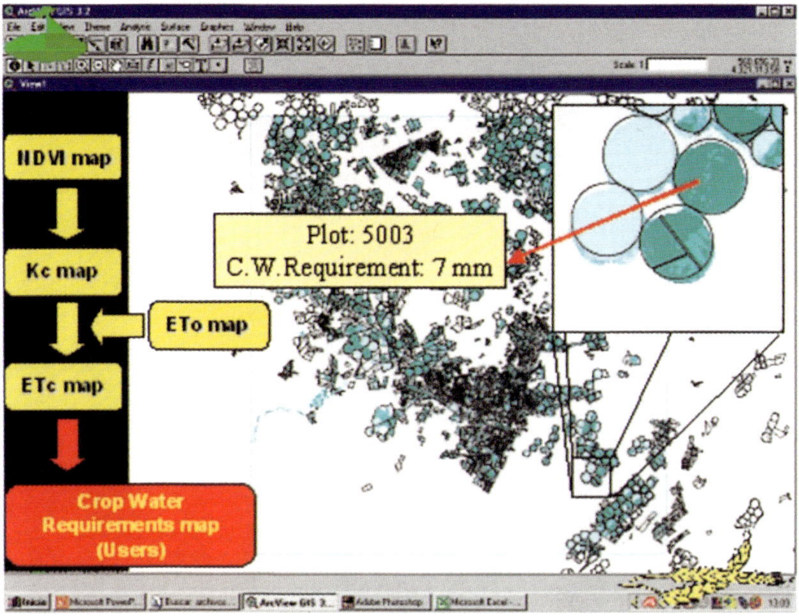

Figure 19.2. *Diagram for crop water requirements derived from the EO data source*

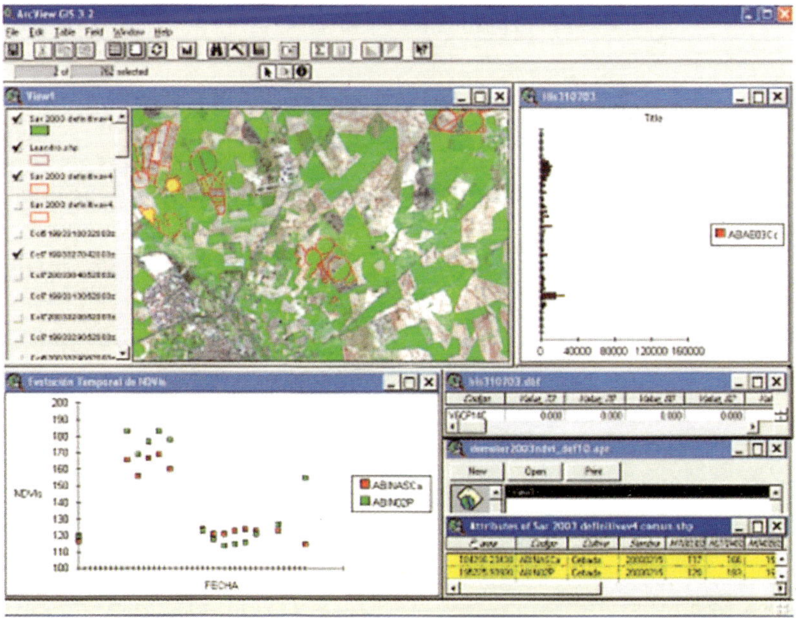

Figure 19.3. *Methodology for Kc Calculation using GIS and image processing technologies*

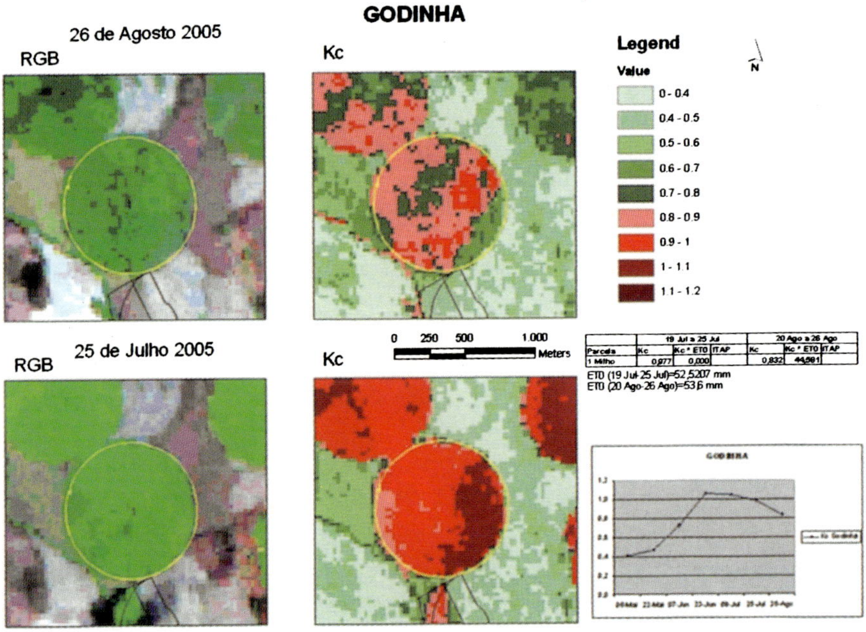

Figure 19.4. *Kc farmer report during the 2005 campaign in the Caia pilot area*

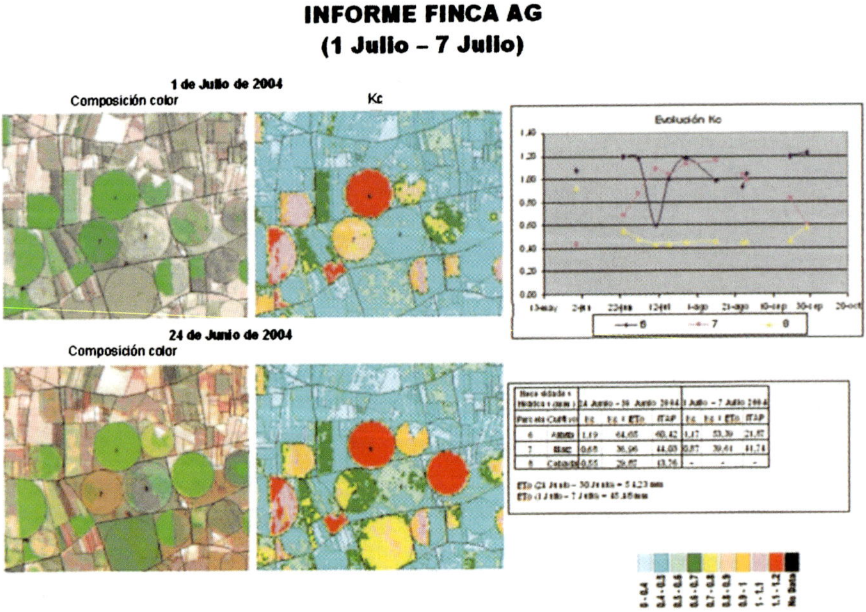

Figure 19.5. *Example of farmer's report distributed weekly to farmers' test groups in Spain*

In this application, DEMO B, other meteorological elements and interpolation methods can be integrated. Actually, this application in a new version was transferred also to ArcGIS.

DEMO C demonstrated the application of a new road ice prediction software across the two European Union Member States Poland and Slovenia of COST719 [CAV 03, DOL 06]. This software IceMiser is a commercial extension based on ArcGIS and combines GIS data with forecast meteorological data in order to produce a 24h forecast of the road surface temperatures network at temporal and spatial resolutions of 20 minutes and 20 meters respectively [CHA 01, CAV 03]. Several GIS layers are required for the application of IceMiser and include Digital Elevation Model (DEM), Cold Air Pooling algorithm, aspect, slope, land use, the road network and the Sky View Factor as the measured point data.

The fourth project, DEMO D, was a Summer School on the Application of GIS in Meteorology and Climatology hosted by the Institute of Biometeorology of the National Research Council (IBIMET-CNR) of Florence (Italy) from 26 to 30 September 2005. The general objective was to reach young scientists in the COST network of the European Institutions in order to transfer state-of-the-art skills and knowledge on the use of GIS for meteorological and climatological applications. Each session included lectures, exercises and discussions and was based on active student participation [DEF 06].

However, due to the fast progress in the development of GIS software and applications, such overviews and classifications including the DEMO projects and the training program have to be adapted to the actual state of the art from time to time.

11.3. GIS applications in meteorology and climatology and regional ecological modeling approaches

This section focuses on user needs and requirements for such GIS applications from the point of view of spatial modelers like ecologists or hydrologists. Ecological modeling approaches such as:

– agroecosystem and forest ecosystem models for the estimation of the impact of different management practices and climatological background conditions on yield, biomass development, water balance and nutrient cycling;

– hydrological catchment models for the estimation of spatial distributed water balance components and discharge rates;

need meteorological data and other inputs such as:

– precipitation, air temperature, air humidity, wind speed, global radiation and soil temperature;

– vegetation data such as type of crop or forest, management data (seeding, harvest etc.);

– soil data such as hydraulic and chemical properties (e.g. hydraulic conductivity, field capacity, wilting point, pH, etc.).

The output of such models consists of:

– transpiration, interception, evaporation;

– surface runoff, ground water recharge, discharge;

– soil water storage (fluxes, water contents, pressure heads);

– nutrient cycling (e.g. N-mineralization, NO_3-leaching, carbon storage);

– crop or forest growth.

Beneath other data, all these models need time series of meteorological input data at the local scale from meteorological stations and also spatial distributions at the regional scale. Spatially distributed accurate estimates of meteorological data are important inputs to such models and are also critical to the model performance. Therefore, a scientific exchange between such modelers and meteorologists is necessary for that purpose and will partly be addressed in this chapter. At first, this leads to methods for quality checks of meteorological time series.

11.3.1. *Quality check of meteorological data*

Meteorological data shows some systematic errors such as the well known underestimation trend in measuring precipitation rates. To take into account these errors in precipitation measurements, there exist a large amount of studies dealing with appropriate correction methods (for example, [SEV 85]). Furthermore, errors in electronic sensors such as those used for the measurement of global radiation can occur. An example of such an error is presented in Figure 11.1.

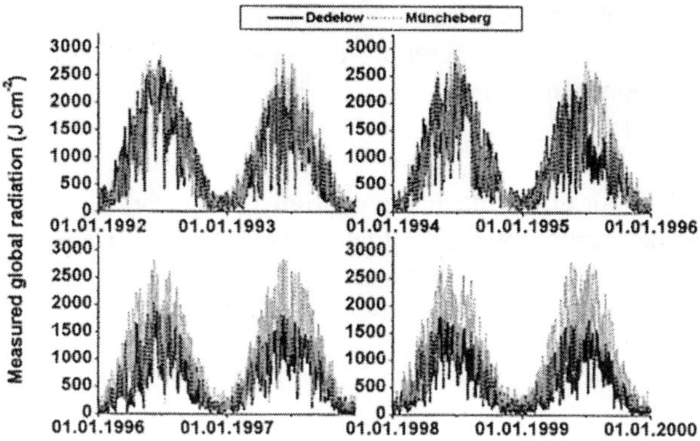

Figure 11.1. *Comparison of daily rates of global radiation in J/cm² measured at two different meteorological stations, 1992-2000*

This figure presents daily rates of global radiation measured at the station Dedelow (latitude: 53.1, elevation: 49 masl) and at Müncheberg (latitude; 52.5 elevation: 62 masl). Both stations are operated by the Center of Agricultural Landscape Research, Müncheberg (ZALF) and are located in the Federal State of Brandenburg, Northeast Germany. The North-South distance between both stations is about 175 km. In July 1996, the station Dedelow started to show distinct under-estimations of global radiation in comparison with the measurements obtained from Müncheberg until the end of the year 2000 (Figure 11.1). After that time, the sensor for global radiation at Dedelow was changed. This example demonstrates the need for a quality check, especially of time series recorded by automatic electronic meteorological sensors.

Another problem is the fact that meteorological time series have gaps in some cases. For the purpose of gaps closure, the simplest way is the direct use of the data of another station. Another method is the application of linear regression lines between two neighbored stations defined as:

$$y_i = a + b \cdot x_i \text{ with } r^2 > 0.7 \text{ and } b > 0.7 \text{ and } < 1.3 \qquad [11.1]$$

where yi represents the missing value from the station with the gaps in the time series and xi is the existent value from the neighbored station with the complete time series. Another possibility is the use of geomathematical models such as combined interpolation and smoothing as well as neural networks.

This indicates the necessity of a further exchange of expert knowledge between meteorologists, climatologists and hydrological as well as ecological modelers at the local as well as at the regional scale. Errors in the meteorological data can lead to errors in the model outputs. Therefore, continuous and precise time series of the meteorological input data are an essential basis for the application of ecosystem models at the local scale.

As an example for the results obtained from an application of a forest ecosystem model at six experimental forested test sites, comparisons of simulated and observed soil water contents are presented in Figure 11.2. These test sites with pine forests are located in the Northeastern German Lowlands in the Federal State of Brandenburg. These forested test sites are located on poor sandy soils with low soil water storage capacity and high infiltration capacity. Beneath meteorological data, the measurements carried out at these test sites include continuous daily measurements of soil water contents and pressure heads with automatic recording time domain reflectometry (TDR) probes and tensiometers from the years 1997-1999.

The age of the pine forests at these six experimental test sites ranges from 80 to 140 years. More information about the forest ecosystem model, the modeling results and the test sites can be obtained from [WEG 03], [RIE 01] and [KAL 01].

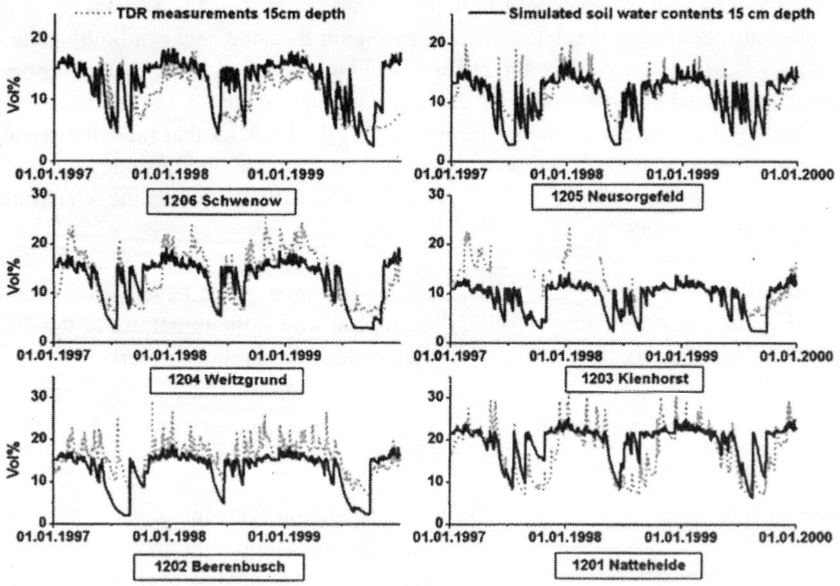

Figure 11.2. *Comparison between simulated and measured soil water contents at the six experimental test sites with pine forest, Brandenburg, NE Germany, 1997-1999 [WEG 03, modified]*

11.3.2. *Regional application of an ecological model*

In order to illustrate the need for the use of appropriate spatial interpolation methods for meteorological data for the application of spatial distributed models, the results of a case study dealing with the application of a simple semi-distributed hydrological catchment model are presented [WEG 06].

The database for the model application consists of:

– Daily precipitation rates, mean wind speed, saturation deficit of air, mean air temperature and global radiation from four meteorological stations and five precipitation stations from 1993 to 2001 operated by Germany's National Meteorological Service (DWD) and ZALF.

– Daily discharge rates from the Pasewalk gauges (1994-2001) located at the Ucker river and Löcknitz (2000-2001) located at the Randow river, both operated by Federal Agency for Environment and Nature Protection Mecklenburg-Vorpommern, Rostock.

– Soil map.

– Land cover data.

– DEM.

The location of the Ucker catchment can be obtained from Figure 11.3. The area of the Ucker catchment is about 2,415 km^2.

Figure 11.3. *Location of the Ucker catchment [WEG 06]*

The areas located in the north of the Ucker catchment are covered with sandy soils with high infiltration rates and low soil water storage capacities. In the southern parts and at the end moraines at higher elevations in the western part of the catchment are located soils with a more loamy texture. In the flat rivers plains, wetlands and peat soils with high ground water tables dominate (Figure 11.4).

The DEM with a grid size of 50 m x 50 m was obtained from the survey offices of Brandenburg and Mecklenburg-Vorpommern. The highest elevation of the catchment with about 145 masl is located in the western part and the lowest one with 0 masl at the mouth of the Ucker into the Stettiner Haff (Figures 11.3 and 11.4). The location of the weather stations and the two gauges can be obtained from Figure 11.4.

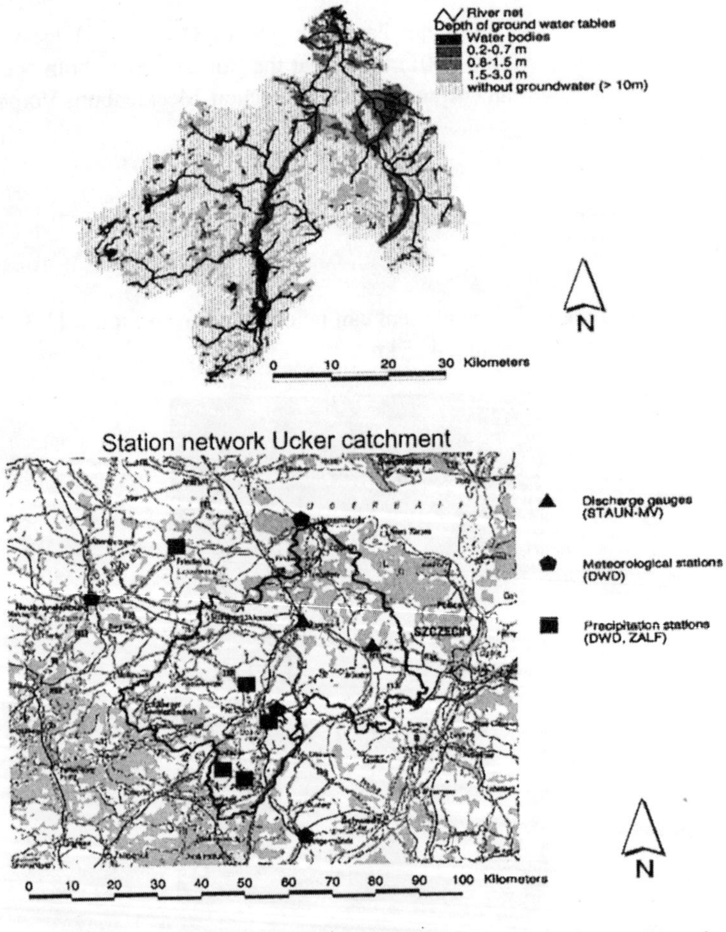

Figure 11.4. *Rivernet and depth of groundwater table (upper graph) and station network (lower graph) of the Ucker catchment [WEG 06]*

Spatial discretization	Hydrotopes, sub-basins, river sections
Time discretization	1 day
Characterization	Conceptual, semi-distributed
Evapotranspiration	HAUDE, TURC-WENDLING, PENMAN-MONTEITH
Vegetation cover	Semi-empirical plant modeling approach
Interception	Single linear storage
Snowmelt	Degree-day approach
Infiltration	Modified HOLTAN approach
Overland channel flow	Linear storage routing of river sections
Unsaturated zone	Multiple layer capacitance model
Saturated zone	Single linear storage – routed over sub-basins

Table 11.1. *General characteristics of the hydrological catchment model [WEG 06]*

An overview of the hydrological catchment model used in our study is given in Table 11.1. The model is coupled to a GIS and needs three coverages for the spatially distributed calculation of water balance and discharge of catchments:

– Hydrotopes: this coverage results from the overlay of the land cover map, the soil map with groundwater level categories as well as the sub-basins map and consists of information about soil type, land cover class or crop type, corresponding sub-basin and mean elevation in the GIS database.

– Sub-basins: this coverage was delineated from the DEM by using GIS operations and consists of information about the mean elevation and the subsoil of each sub-basin.

– Rivernet (Figure 11.4): this coverage results from the overlay of the sub-basins map with the DEM and the rivernet. Furthermore, this dataset includes information about the cross-profile and the elevation of the river nodes as well as a reference to the corresponding downstream river section.

For each hydrotope, water balance and runoff is calculated on daily time steps. The daily rate of potential reference evapotranspiration PET is calculated according to [WEN 91] as a function of global radiation, saturation deficit of air and wind speed. The partitioning of PET in transpiration, interception and evaporation is estimated by a semi-empirical plant model by using simple crop specific seasonal time courses for rooting depth, plant height and soil cover [KOI 90]. The infiltration of throughfall (precipitation minus interception) into the topsoil layer and the generation of surface runoff is calculated by a modified, semi-empirical approach according to [HOL 61]. The actual soil water content and the water fluxes of each soil layer are calculated by using a plate theory approach with a non-linear storage routing technique according to [GLU 69]. The water flux across the lower boundary of the soil profile is defined as a groundwater recharge. In the presence of groundwater in the soil profile, a capillary rise is calculated according to the German Soil Classification rules [AGB 94] depending on soil texture, bulk density, distance from the soil layer to groundwater table and soil water content in the layer. The soil type of each hydrotope in the GIS database is described by a corresponding representative 1D vertical soil profile. Based on the texture and bulk density of each soil layer, which were obtained from these representative soil profiles, soil parameters like field capacity and wilting point for the soil water model were evaluated for each layer of the corresponding soil profile according to the German Soil Classification rules [AGB 94].

For the modeling time step, surface runoff and groundwater recharge rates are calculated separately for each single hydrotope and accumulated for the corresponding sub-basin for the purpose of stream flow routing. For each sub-basin, the accumulated surface runoff is routed by a fast linear storage and the accumulated groundwater recharge is routed by a slow linear storage for the base flow. The stream flow through the rivernet is simulated by linear storage cascades [WEG 06]. Land cover parameters like soil cover and rooting depth for each land cover or crop type in the GIS database are also stored in a separate vegetation data table.

For the simulation period from 1st January 1996 to 31st December 2001 presented in our study, the spatial distributions of daily values of precipitation, air temperature, saturation deficit of air, wind speed and solar radiation obtained from four meteorological stations and five precipitation stations were estimated by using a simple inverse distance weighted interpolation procedure included in the model [WEG 06]. Measured daily precipitation rates are in most cases too low, in the order of about 5-15% (for example, [SEV 85]). Therefore, daily rates of measured precipitation were corrected within the model by multiplying the precipitation rates with a factor of 1.1. The initial soil water contents in the catchment for the simulations runs of the model were set equal to field capacity.

An example of the results of the interpolation of meteorological input data for two days with different daily rates of precipitation in the simulation period is presented in Figure 11.5.

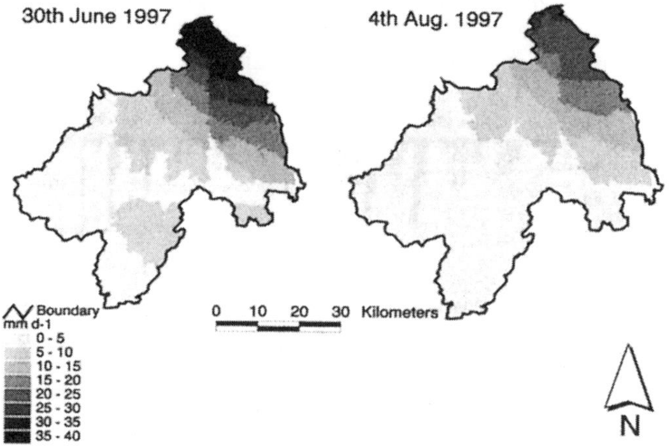

Figure 11.5. *Spatial distributed daily rates of precipitation for two days interpolated by the inverse distance weighted interpolation procedure within the model runs*

These spatial distributions of the meteorological input data such as those presented in Figure 11.5 have a strong impact on the results of a hydrological catchment model such as those presented in Figures 11.6 and 11.7. For the spatialization there are two modeling strategies. In the first strategy the meteorological input data is interpolated outside the hydrological model by using appropriate GIS tools for the interpolation such as Demo B. After that preprocessing, the spatial distributions of the data are provided as daily input files for the simulation runs of the model. The second strategy, like that in our study, uses a spatial interpolator which is integrated in the simulation model and the interpolation procedure at each modeling time step is carried out within the simulations runs of the model. An additional, more detailed description of the GIS database and the model can be obtained from [WEG 06].

11.4. Conclusions

Hydrological and ecological modelers are confronted with a variety of stochastic and deterministic methods such as inverse distance weighted averaging, splining, polynomial regression, trend surface analysis, kriging and cokriging for the spatial interpolation of meteorological data. Most of those applying the model, such as ecologists, geographers and hydrologists, often have only basic meteorological knowledge and, therefore need advice, education and training in the use of GIS, interpolation methods and appropriate as well as effective software tools to process their meteorological input data for spatial model applications from meteorological research groups such as COST719. This is an essential basis for the simulation quality of such models at a local and regional scale.

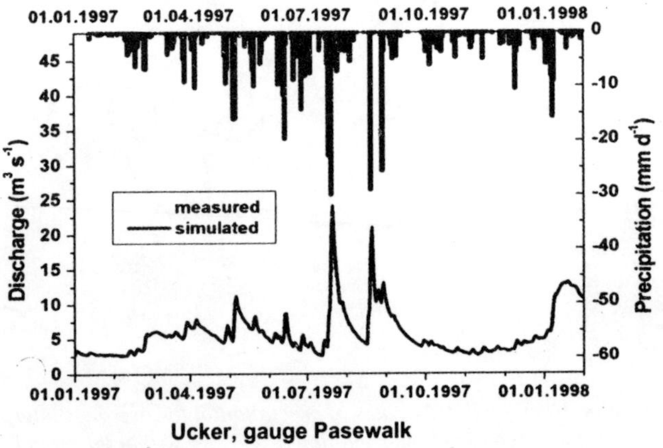

Figure 11.6. *Daily rates of precipitation and simulated and observed discharge in 1997*

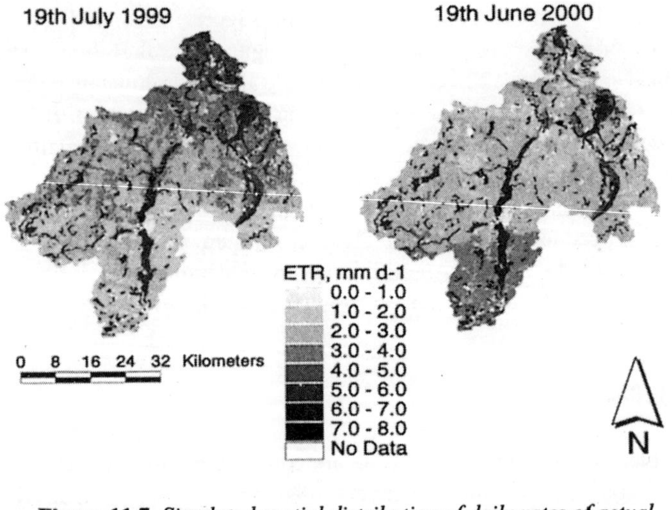

Figure 11.7. *Simulated spatial distribution of daily rates of actual evapotranspiration (ETR) for two days in the Ucker catchment*

11.5. Acknowledgements

The case study in the Ucker catchment was founded by the German Federal Ministry of Consumer Protection, Food and Agriculture (BMVEL) and the Ministry of Agriculture, Environmental Protection and Landscape Planning (MLUR) in Brandenburg. We have to thank the German Meteorological Service for providing meteorological data for the Ucker catchment. Additionally, we have to thank P. Einert, R. Kallweit, A. Konopatzky, W. Riek and B. Ströhbach from the Federal Forest Agency Eberswalde in Brandenburg for providing the data from the six experimental field plots with pine forest in Brandenburg. We also have to thank the Federal Agency for Environment and Nature Protection Mecklenburg-Vorpommern and the Agency for Environment, Nature protection and Geology Mecklenburg-Vorpommern for providing discharge data for the Ucker catchment. The digital elevation model DEM50 of the Ucker region was provided by the Federal Agencies Landesvermessungsamt Brandenburg and Landesvermessungsamt Mecklenburg-Vorpommern.

11.6. Bibliography

[AGB 94] AG BODEN, Bodenkundliche Kartieranleitung, 4. Aufl. Schweizerbart Stuttgart, 1994.

[CHA 01] CHAPMAN L., THORNES J.E., BRADLEY A.V., "Modelling of road surface temperature from a geographical parameter database. Part 2: Numerical Analyses", Meteorological Applications 8, 2001, p. 421-436.

[CHA 03] CHAPMAN L., THORNES J.E., "The use of geographical information systems in climatology and meteorology", Progress in Physical Geography 27, 2003, p. 313-330.

[CAV 03] CAVAN G., CHAPMAN L., THORNES J.E., The use of GIS and IceMiser to predict winter road surface temperatures in Poland, Cost 719 WG3 Working paper: STSM Report, 2003.

[DEF 06] DE FILIPPIS T., DI VECCHIA A., MARACCHI G., SORANI F., "Training programme for the dissemination of climatological and meteorological applications using GIS technology", Adv. Geosc. 8, 2006, p. 19-26.

[DOL 07] DOLINAR M., KENDA K., BUH S., The use of GIS and "IceMiser" to predict winter road surface temperatures in Slovenia.- COST 719 End report (in press), 2007.

[DYR 05a] DYRAS I., DOBESCH H., GRUETER E., PERDIGAO A., TVEITO O.E., THORNES J.E., VAN DER WEL F., BOTTIA I., "The use of geographic information systems in climatology and meteorology: COST 719", Meteorological Applications Vol. 12, No. 1, 2005a, p. 1-7.

[DYR 05b] DYRAS I., SERAFIN-REK D., "The application of GIS-technology for precipitation mapping", Meteorological Applications Vol. 12, No. 1, 2005b, p. 69-77.

[GLU 69] GLUGLA G., "Berechnungsverfahren zur Ermittlung des aktuellen Wassergehaltes und des Gravitationsabflusses", Albrecht-Thaer-Archiv 13, 1969, p. 371-376.

[HOL 61] HOLTAN H.N., "A concept for infiltration estimates in watershed engineering", US Dept. ARS, 1961, p. 41-51.

[KAL 01] KALLWEIT R., "Kalkulation von Vorräten und Umsätzen oberirdischer Biomassekompartimente und Nährstoffmengen der Kiefernbestände a Level-II-Flächen Brandenburgs", Beiträge für Forstwirtschaft und Landschaftsökologie 35, 2001, p. 37-44.

[KOI 90] KOITZSCH R., GÜNTHER R., "Modell zur ganzjährigen Simulation der Verdunstung und der Bodenfeuchte landwirtschaftlicher Nutzflächen", Arch. Acker-Pflanzenbau Bodenkd. 24, 1990, p. 717-725.

[RIE 01] RIEK W., STROHBACH B., "Untersuchungen zum Wasserhaushalt auf Level II-Standorten in Brandenburg", in: Landesforstanstalt Eberswalde (Hrsg.): Forstliche Umweltkontrolle – Ergebnisse aus zehnjährigen Untersuchungen zur Wirkung von Luftverunreinigungen in Brandenburgs Wäldern. Eberswalde, Potsdam, Hendrik Bäßler Verlag, Berlin, 2001, p. 116-131.

[SEV 85] SEVRUK B., "Correction of precipitation measurements, summary report", Proceedings of the ETH/IAHS/WMO Workshop of the Correction of Precipitation Data, 1985, p. 13-23.

[WEG 03] WEGEHENKEL M., JOCHHEIM H., "Modellierung des Wasserhaushaltes von Kiefernbeständen des Level-II-Programms in Brandenburg mit unterschiedlich komplexen Simulationsmodellen", Forstw. Cbl. 122, 2003, p. 302-317.

[WEG 06] WEGEHENKEL M., HEINRICH U., UHLEMANN ST., DUNGER V., MATSCHULLAT J., "The impact of different spatial land cover datasets on the outputs of a hydrological model – a modelling exercise in the Ucker catchment, North-East Germany", Physics and Chemistry of the Earth, 2006, 31(17), p. 1075-1108 .

[WEL 05] VAN DER WEL F.J.M., "Spatial data infrastructure for meteorological a climatic data", Meteorological Applications, Vol. 12, No. 1, 2005, p. 7-9.

[WEN 91] WENDLING U.H., SCHELLIN G., THOMÄ M., "Bereitstellung von täglichen Informationen zum Wasserhaushalt des Bodens für die Zwecke der agrarmeteorologischen Beratung", Z. Metereol., 41(6), 1991, p. 1-16.

[TVE 05] TVEITO O.E., BJORDAL I., SKELVAG A.O., AUNE B., "A GIS based agroecological decision system based on gridded climatology", Meteorological Applications Vol. 12, No. 1, 2005, p. 57-69.

Chapter 12

GIS Application to Daily Fire Risk Mapping

12.1. Introduction

Forest fires in Portugal cause each year great economic, environmental and in some cases life losses.

Since the summer of 2004 the Geographic Information Systems (GIS) has been used in the Portuguese Meteorological Institute to map the daily fire risk in Portugal mainland [IMP 06a]. However, the Fire Weather Index (FWI), which is one of its components, was originally developed by the Canadian Forest Service [CFS 06] and has been calculated and spread since 1998 [IMP 06b].

The fire risk map consists of merging the conjuncture and the weather indices. Due to the operational organization of the Civil Protection System the information of the fire risk by administrative regions is very useful, since in this way the grid of the fire risk map is used to calculate the spatial statistics for the several regions.

The GIS is used in the mapping process from the interpolation, passing by the map algebra, until the layout creation.

All the maps are produced in real-time in order to provide the information as soon as possible. The automation in GIS has been done in ArcGIS using Visual Basic and ArcObjects [ESR 06].

Chapter written by Álvaro SILVA.

12.2. Methodology

The fire risk map is obtained from the combination of the conjuncture and the weather components.

12.2.1. *Conjuncture Fire Index (CFI)*

The conjuncture component (CFI) combines historical information on the number of fires, burnt areas, type of vegetation, biomass accumulation and climatological variables. A lot of this information was obtained by remote sensing [PER 05]. The result is a grid for the 2006 fire season (see Figure 12.2) and what varies daily in the fire risk map is the FWI.

The CFI varies from 1, which is the lowest risk, in the coastal areas and in some southern regions, to 5, which is the maximum risk, spread all over the Northern and Central regions and in some mountainous areas of the Algarve in the South.

12.2.2. *Fire Weather Index (FWI)*

The weather component that is the Fire Weather Index (FWI) combines several sub-indices like the Fine Fuel Moisture Content (FFMC), Duff Moisture Code (DMC), Drought Code (DC), Initial Spread Index (ISI) and the Build-Up Index (BUI) (see Figure 12.1), which express different types of information:

– "the FFMC classifies the fine dead fuels, with rapid drying rate, according to their moisture content. Therefore, it corresponds to the degree of flammability of these fuels, which are on the soil surface. The moisture content of these fuels at 12 UTC on a given day depends on the moisture content at the same hour on the previous day, the precipitation (mm) occurred in 24 hours (12-12 UTC) and the air temperature (°C) and relative humidity (%) at 12 UTC on the day considered. The wind speed only influences the drying rate of these materials;

– the DMC expresses the moisture content of humus and woody materials of medium size up to nearly 8 cm under the soil surface. The humus index is estimated from the precipitation occurred in 24 hours (12-12 UTC), the air temperature and relative humidity at 12 UTC and the humus index of the previous days;

– the DC is a good indicator of the effects of seasonal drought on forest fuels (humus and woody materials of larger size), which are under the soil surface, from 8 to 20 cm deep. This drought index is obtained from the precipitation occurred in 24 hours, the temperature at 12 UTC and the previous day DC;

– the ISI focuses the initial spread of the fire that depends on the FFMC sub-index and the wind speed (km/h) at 12 UTC;

– finally, the BUI is a factor of evaluation of the vegetables that can build up a fire ("heavy" fuels on the soil) and is estimated from two of the sub-indices: DMC and DC" [IMP 06b].

These indices are calculated daily, at 12 UTC hours, by using the air temperature, the amount of precipitation occurred over the past 24 hours, the wind speed and the relative humidity measurements at the automatic network stations. The resulting FWI is not only a synthesis of these values and sub-indices for a given day but also take into account the evolution of conditions and therefore it can be considered a cumulative index.

The forecast for the next 24 and 48 hours is also produced in the same way but the observations at the stations are substituted with the run of the mesoscale model ALADIN.

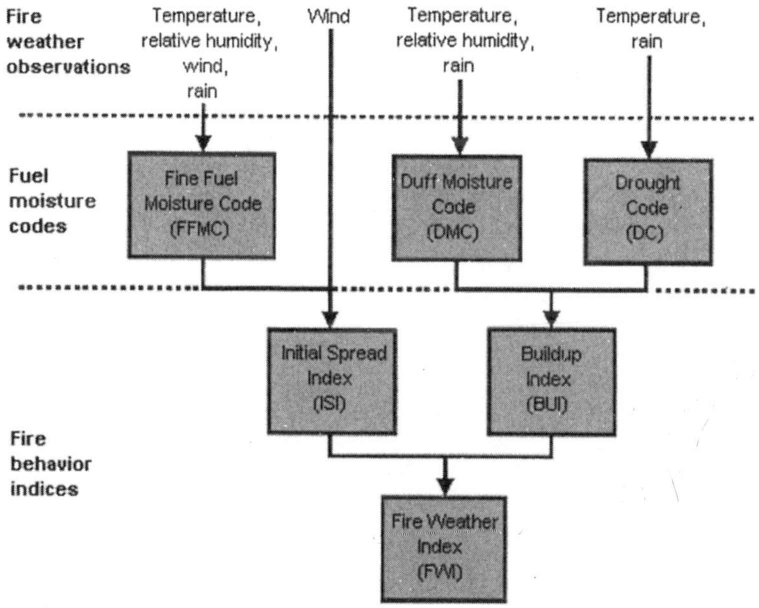

Figure 12.1. *Fire weather index components [CFS 06]*

12.2.3. *Fire risk mapping*

The GIS application to the fire risk has been done in ArcMap 8.3 with Spatial Analyst by using the VB scripting language and ArcObjects.

To simplify the automation procedures, a quick but reliable interpolation method was chosen; the Inverse Distance Weighting (IDW), which is widely known and applied, was tested for some typical days presenting acceptable results and it is used in real-time to obtain the grid of FWI.

The grids of CFI and FWI are merged to obtain the Combined Fire Risk (CFR) that varies from 1 (lowest risk) to 5 (highest risk). The hottest, driest and windy regions should present the highest FWI values but not necessarily the maximum risk values in the final CFR map (Figure 12.3); the weather conditions can be suitable to originate a large fire if an ignition occurs, but if there is no vegetation in that area the risk will be low and so on. This kind of combinations between the two indices is addressed in a spatial logical function.

12.3. Results: some examples

12.3.1. *FWI map*

The example of the fire risk map for 6th September (see Figure 12.3) shows a typical summer day with very high FWI values all over the country with the exception of the coastal regions. The lack of precipitation in the last weeks and the high temperatures are responsible for a very dry soil with the fuels presenting a low level of moisture. It can be said that 2/3 of the country exhibits the maximum fire weather risk.

Despite this not all the regions will be classified with the highest combined risk as described in the next section.

12.3.2. *CFR map*

The CFR map is the main output for fire prevention and combat because it combines the weather and the conjuncture indices as mentioned above. A high correlation can be seen (see Figure 12.4) between the highest FWI values (see Figure 12.3) and the maximum CFI class (see Figure 12.1) in the continental Northern and Central regions but not in the main part of the Southern regions. These present the lowest CRI class so, in the end, despite the high FWI values, they are classified risk 2 or moderate.

The opposite is not so clear in this example, namely due to the fact that almost all the country presents high weather risk. However, if we take a closer look we can see, for instance, some areas classified 4 in the conjuncture index which appear 5 in the combined risk, thus exemplifying one of the many conditions: if one of the indices has the highest risk then the CFR will be also the maximum.

12.3.3. *Adding spatial statistics*

In the GIS program also the zonal statistics are calculated because many times the Civil Protection System needs the mean value of risk for administrative regions (see Figure 12.5). The tabular information for these regions is also useful considering that there are some regions with considerable differences within, which is reflected in the square deviation of the fire risk values, for example. Joining this information to the mean fire risk map can help identifying critical areas where the mean is not representative and it is necessary to take a look at the CFR map.

12.4. Conclusion

This chapter showed an experience in using the GIS to operational map daily fire risk. Processes were expressed in a very practical way focusing on the application itself rather than large theoretical assumptions. Further developments are still needed and will be carried out. At the moment GIS is used for:

– interpolation of the FWI values;

– map algebra – merging the conjuncture index and the weather index;

– zonal statistics and calculation of the statistical measure for each region;

– production of visually appealing and easy to understand layouts;

– geodatabase of georeferenced information concerning the fire risk.

In the future GIS will be used to:

– put all the processes in GIS;

– focus on the interpolation of the basic meteorological elcments;

– access uncertainty.

The application of GIS in merging meteorology and environmental matters, as stated in the described fire risk example, will, in the author's point of view, increase in the near future. Also the climatic change relations with biological, medical and social aspects are already a reality and the GIS can play a significant role in it.

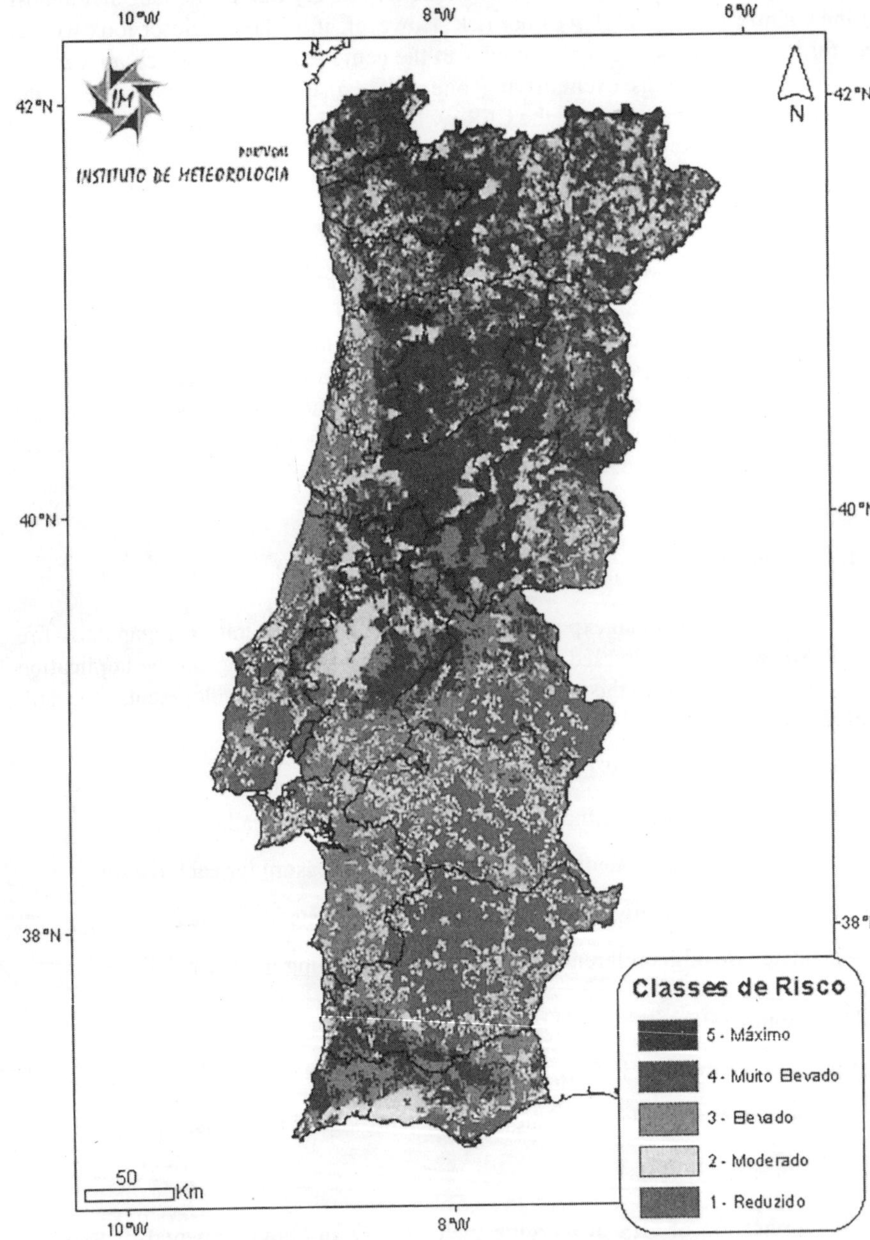

Figure 12.2. *CFI for Portugal mainland in 2006 [ISA 06]*

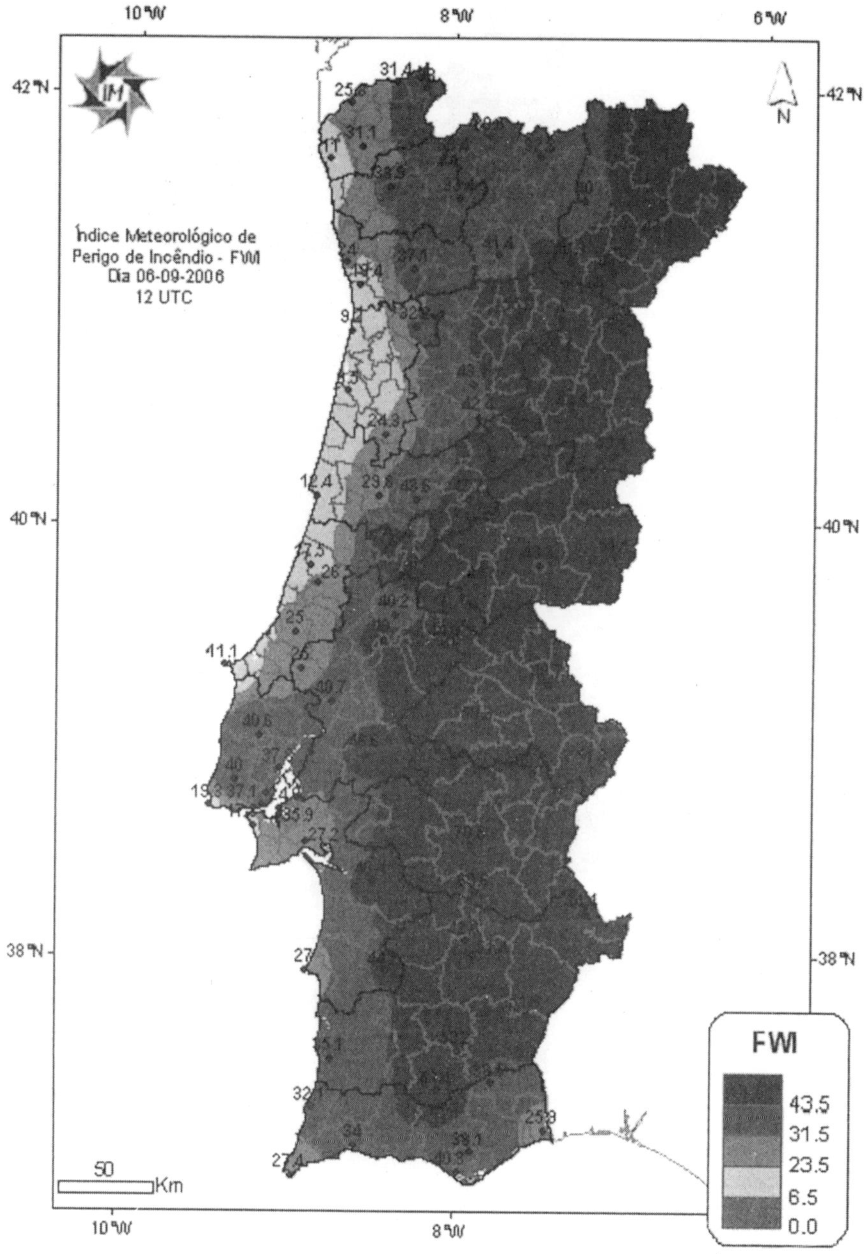

Figure 12.3. *FWI for Portugal mainland, 6th September 2006 (the labeled points represent the FWI values at the meteorological stations)*

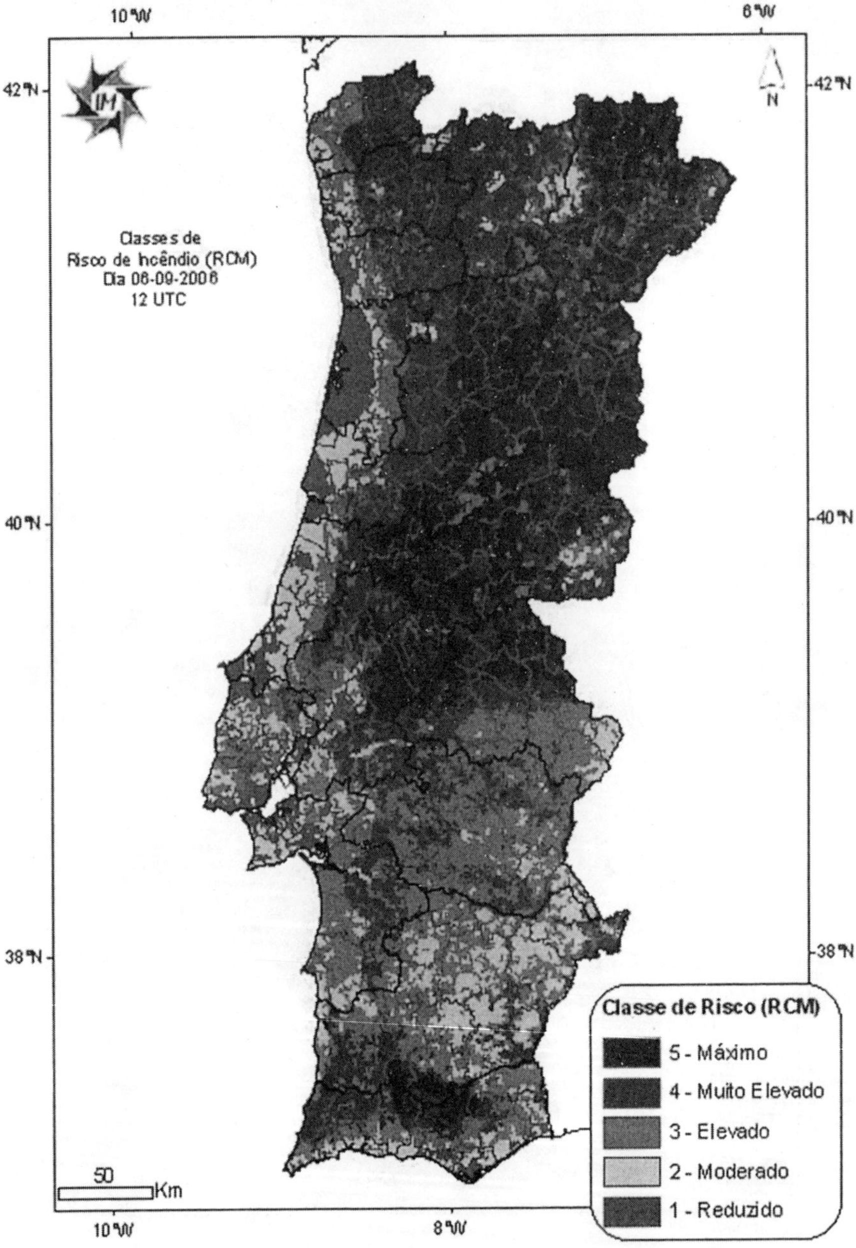

Figure 12.4. *CFR for Portugal mainland, 6th September 2006*

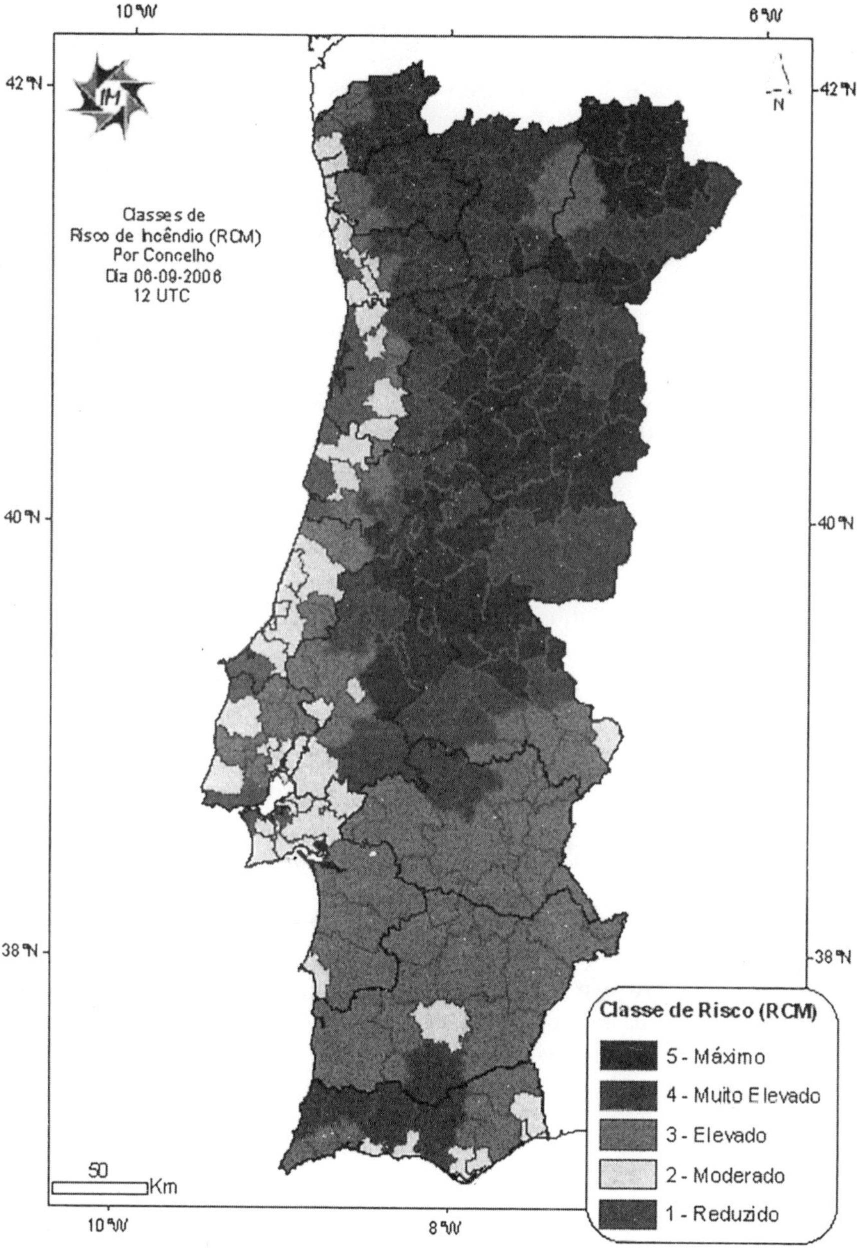

Figure 12.5. *CFR for the municipality level in Portugal mainland, 6th September 2006*

12.5. Bibliography

[BUR 00] BURROUGH P., MCDONNELL R., Principles of Geographical Information Systems, New York, USA, Oxford University Press, 2000.

[CFS 06] CANADIAN FOREST SERVICE, Canadian Forest Fire Weather Index (FWI) System, http://cwfis.cfs.nrcan.gc.ca/en/background/bi_FWI_summary_e.php.

[DUB 03] DUBOIS G., MALCZEWSKI J., CORT M., Mapping Radioactivity in the environment. Spatial Interpolation Comparison 97, Luxembourg, Office for Official Publications of the European Communities, 2003.

[ESR 06] Environmental Systems Research Institute, Inc. (ESRI), ArcObjects Technology, http://www.esri.com/software/arcgis/about/arcobjects.html.

[IMP 06a] Portuguese Meteorological Institute, Meteorological Support to Forest Fire Prevention – Fire Risk, http://web.meteo.pt/en/previsao/riscoincendio/prev_risc_class_conc.

[IMP 06b] Portuguese Meteorological Institute, Fire Risk Index (FWI), http://web.meteo.pt/en/previsao/riscoincendio/risc_incend_fwi.html.

[ISA 06] Instituto Superior de Agronomia for Direcção Geral dos Recursos Florestais, Carta de Risco de Incêndio Conjuntural, Lisbon, 2006.

[JOH 01] JOHNSTON K. *et al.,* Using ArcGIS Geostatistical Analyst, Redlands, USA, ESRI, 2001.

[PER 05] PEREIRA J., CARREIRAS J., Carta de Risco Conjuntural de Incêndio Florestal (CRCIF) v. 2005, http://www.adetti.pt/events/PIMHAI05/files/JMCPereira.pdf.

[RAF 05] RAFFUSE S., BROWN S., CHINKIN L., SULLIVAN D., "Estimating Regional Contributions to Atmospheric Haze Using GIS", ESRI International User Conference Proceedings, USA, 2005.

[SIL 04] SILVA A., SOUSA A., ESPÍRITO SANTO F., "Mean air temperature estimation in Mainland Portugal: Test and comparison of spatial interpolation methods in GIS", in Proceedings of Conference on the Spatial Interpolation Techniques in the Climatology and Meteorology, Budapest, Hungary, 2004.

[SIJ 06] SILVA J., GOMES P., Índice Meteorológico de Risco de Incêndio Florestal – FWI e Classes de Risco de Incêndio – Relatório Mensal Agosto 2006, http://web.meteo.pt/resources/im/pdfs/fogos_rc_08_06.pdf.

Chapter 13

Application of GIS Technology on the Comparisons of Climatological Databases: An Overview of Winter Precipitation over Spain

13.1. Introduction

Nowadays, there is a substantial and growing interest in new technologies that makes it possible to improve the framework for managing and integrating datasets. In that sense, Geographical Information Systems (GIS) constitute a computer technology that uses geographic information as an analytic framework, solving problems associated with the organization and combination of data. Concerning climate, a GIS is currently associated with maps, providing rather great capabilities to resolve problems than using a simple mapping program. In that particular way, the generation of an Iberian daily precipitation database by means of statistical spatial interpolation of *in-situ* measurements onto a regular grid by using the GIS has been one of the objectives of the Spanish Meteorological Service (Instituto Nacional de Meteorología, INM). As a consequence of these efforts, INM has produced a complete high-resolution long-term precipitation database over the Iberian Peninsula and the Balearic Islands. This dataset became a rather powerful tool to be applied in different climatic studies.

Chapter written by M.Y. LUNA, M.L. MARTÍN, M.G. SOTILLO, C. ALMARZA, F. VALERO and J. DE LA CRUZ.

There has recently been a growing concern about the need for long-term information about prevailing environmental conditions. In that sense, the generation by Puertos del Estado (PE) within the HIPOCAS (Hindcast of Dynamic Processes of the Ocean and Coastal Areas of Europe) Project framework of a Mediterranean long-term (1958-2001) homogenous high resolution atmospheric database provides an useful dataset for regional studies. The already proved reliability of the hindcasted database [RAT 03] and [SOT 06] enhances the confidence on the HIPOCAS products. However, concerning precipitation, it would be needed to perform a comparison between the hindcasted data and a good observation-based dataset. In order to fulfill this objective, this chapter shows an exhaustive comparison between the HIPOCAS precipitation and the comprehensive INM GIS-constructed precipitation database over the Iberian Peninsula. Additionally, an evaluation of the potential improvement of the HIPOCAS data versus other reanalyzed data, such as the NCEP and ERA datasets, are shown. All the products needed to carry out the comparative analysis were calculated, digitalized and depicted by means of GIS tools, making this work a direct application of GIS technology to climatic studies. The chapter is organized as follows. There is a brief description of the INM GIS-constructed Iberian precipitation database along with the HIPOCAS and global reanalysis precipitation datasets in section 13.2. This section is also devoted to the description of the methodology used. Section 13.3 presents the results derived from the comparisons of the mentioned datasets and finally, the main conclusions are drawn in section 13.4.

13.2. Data and methodology

A high-resolution daily precipitation database was generated by the INM by using *in-situ* measurements from the INM station network. The INM Climatological Area has elaborated an Iberian daily precipitation database by means of statistical spatial interpolation of *in-situ* measurements onto a regular grid. The objective was to build a complete daily dataset which is necessary as input of spatially distributed models and for understanding the climate variability on a daily basis. All the *in-situ* measurements of daily precipitation data from the Historical Database of INM were extracted for the period 1 January 1961 to 31 December 2003, regardless of their time coverage. These stations, which are irregularly distributed over the Iberian Peninsula and the Balearic Islands (Figure 13.1a), provided a good spatial coverage over the whole domain. These irregularly spatial distributed *in-situ* measurements were interpolated by the INM onto a 25-km regular grid. The interpolation was performed by using the Kriging method implemented in a GIS software (ESRI® ArcMap 8.3 Software). This interpolation technique preserves more variance than other methods such as the inverse distance weighting method [SHE 01]; additionally, this spatial interpolation tool is widespread in software related to GIS, thus enabling comparisons with databases from different countries. All these considerations are outlined in the COST Action 719 (The Use of GIS in Meteorology and Climatology) in which the INM is an active participant. Further information about this precipitation dataset can be found in [LUN 04].

The HIPOCAS long-term database is the result of an atmospheric hindcast performed over the whole Mediterranean Basin [SOT 06]. In order to produce the 44-year (1958-2001) hindcast the regional atmospheric model REMO was used. The Mediterranean hindcast was performed with a horizontal resolution of 0.5° x 0.5° (roughly 50 x 50 km^2). The model domain (Figure 13.1b) is wide enough to incorporate the whole Mediterranean Basin within the forecast area. Additionally, a spectral nudging technique was applied to keep the regional run close to the imposed time-variable large-scale atmospheric states provided by the NCEP forcing. This chapter is exclusively focused on the analysis of the winter precipitation over the Iberian Peninsula and the Balearic Islands. Thus, the selected domain spans from 10° W to 5° E and from 35° N to 45° N and a HIPOCAS subset (December to February) precipitation was extracted from the whole HIPOCAS Mediterranean domain over the area of interest. The analysis is performed with monthly accumulated precipitation values for a 41-year period, covering from 1st January 1961 to 31st December 2001. In this chapter, the winter HIPOCAS precipitation data over the Iberian Peninsula and the Balearic Islands is compared with the aforementioned observational INM dataset and other climatic datasets. To do this, the winter monthly observed Iberian Precipitation Dataset (IPD) observed over 41 years (1961-2001) was used. Thus, the original INM daily data was interpolated to the HIPOCAS grid and the monthly accumulated precipitation values were obtained. Additionally, monthly precipitation from NCEP/NCAR [KAL 96] and ERA [GIB 97] global reanalyzes were also interpolated to the HIPOCAS grid and were used in order to complete the HIPOCAS validation.

The behavior of the winter HIPOCAS precipitation data over the Iberian Peninsula and the Balearic Islands is analyzed by a general description using several derived products. In addition, the possible improvement of the winter HIPOCAS with respect to the remainder of the databases is performed by means of its comparative analysis against the observational IPD and the global reanalysis. In this chapter, the GIS methodology has been used to calculate and digitalize the statistical patterns associated with those derived products for both observed and simulated precipitation fields because of the easiness of GIS tools to manage and integrate datasets. Here, precipitation means, variances and bias (simulated datum minus observed one) will be used to characterize the winter rainfall regime of both datasets. Additionally, to carry out the comparisons between the different datasets, the root mean squared error (RMSE) and the Relative Bias (RB) are also derived. Finally, the temporal evolution of the spatial average bias and RMSE are derived to evaluate the different model performance ability.

13.3. Results

The 41-winter monthly mean IPD precipitation field (Figure 13.2a) shows the well-known differences in the precipitation behavior between the Atlantic and Mediterranean areas of the Iberian Peninsula. This field is characterized by a strong gradient with a maximum (minimum) located over the northwestern (southeastern) side of the Iberian Peninsula. The constructed mean HIPOCAS precipitation field (Figure 13.2b) agrees with these known differences, not only in the spatial gradient but also in the absolute precipitation values. A similar comparison was performed by using the NCEP and ERA global reanalysis datasets (Figures 13.2c and 13.2d), showing similar spatial distributions and reaching up to 140 mm in the case of ERA and 95 mm for NCEP. These maximum values are clearly lower than the observed IPD and hindcasted HIPOCAS (Figures 13.2a and 13.2b), which are of the order of 220 mm in both cases. Thus, the important improvement in the characterization of the observed precipitation introduced by the HIPOCAS hindcast in relation to the other global reanalyzes is remarkable. The variance patterns (Figure 13.3) have also been constructed by the GIS method, providing an in-depth look into the precipitation variability over time and space in the Iberian Peninsula. Both IPD and HIPOCAS variance fields show similar contour patterns, that is, the higher the IPD variance (Figure 13.3a), the higher the HIPOCAS variance (Figure 13.3b). The analogy between the precipitation means and the variances of IPD and HIPOCAS fields, which is reflected not only in the spatial distribution but also in the absolute precipitation values, highlights the significant improvement introduced by the HIPOCAS hindcast. If these variance patterns are compared with NCEP and ERA variance configurations (Figures 13.3c and 13.3d), it is remarkable in terms of absolute values the great difference for reproducing the observed IPD variability field. While the NCEP and ERA only reach up to 60 mm^2 and 80 mm^2 in the variance fields, respectively, the HIPOCAS database appropriately reproduces the observed variability, showing similar distribution of maximum and minimum values to the IPD variance and reaching up analogous values.

The GIS technique is also used to calculate the bias and relative bias between the IPD set and the remainder of the simulated datasets. The produced bias between the observed and simulated fields (not shown) is in absolute value lower than 20 mm over most of the Iberian Peninsula, showing that the HIPOCAS precipitation field reproduces realistically the observed values. The RB values show manifest and remarkable differences between them (Figure 13.4). The RB associated with the HIPOCAS field (Figure 13.4a) shows most of the study area marked by lower and negative RB, which indicates a general light underestimation by the HIPOCAS database. A zone of negative RB is derived over southwestern Iberia with a regional minimum along the Strait of Gibraltar. It is in this area where the bias is more significant in terms of percentage of total observed precipitation, since the values are comprised between 40% and more than 100%, as noted in Figure 13.4a,

Nevertheless, rather than an unrealistic HIPOCAS simulation, the existence of these biased areas seems to be more linked to a displacement of the hindcasted precipitation maximum (as seen in Figure 13.2b) compared to the observed IPD one (see Figure 13.2a). The RBs associated with the NCEP and ERA fields (Figures 13.4b and 13.4c) show high negative values over all areas, thus exhibiting marked west-east gradients that are not visible in the HIPOCAS-RB. In order to characterize the observed Iberian precipitation field, a more complete view of the NCEP, ERA and HIPOCAS performances is provided. Figure 13.5 is created to display the time evolution of the bias (Figure 13.5a) and the RMSE (Figure 13.5b) averaged over the whole spatial domain. It is worth noting that whereas both reanalyzes underestimate significantly the IPD field, showing a negative bias of the mean precipitation along the entire time period, the HIPOCAS bias fluctuates around zero and does not show any significant underestimation of the observed precipitation. The RMSE of HIPOCAS has shown the lowest errors, highlighting the remarkable improvement introduced by the hindcasted with respect to the current global reanalysis data. Spatial patterns of time correlation between the observed IPD and the simulated (HIPOCAS, NCEP and ERA) fields have also been obtained but the maps are not shown in this chapter. They exhibit a good correlation between them with the highest values over the western Iberia side and the lowest over the Mediterranean coast. The similarity between all the time correlations between the observed dataset and the three simulated datasets indicates that although the absolute values are not well reproduced in both ERA and NCEP data, all databases simulate the time evolution of precipitation in an adequate manner.

13.4. Summary and conclusions

The GIS methodology is used in this chapter to compare and analyze the winter HIPOCAS monthly precipitation with the Iberian winter monthly accumulated precipitation dataset over Spain. Additionally, the HIPOCAS precipitation is compared with the NCEP and ERA reanalyzes in order to provide a broader picture of such data quality. The analysis was based on winter (DJF) monthly-accumulated precipitation values for a 41-year period, spanning from 1st January 1961 to 31st December 2001.

The behavior of the simulated precipitation datasets over Spain has been analyzed by using several GIS-derived products. The statistical comparative analysis performed between the long-term simulated HIPOCAS precipitation field and the observed IPD highlights the good agreement in terms of spatial and temporal distribution, as well as in terms of total amount of precipitation. The mean and variance HIPOCAS precipitation patterns have shown good agreements in comparison with the observed precipitation dataset, not only in the spatial gradient but also in the absolute precipitation values. Although similar NCEP and ERA patterns exhibit similar spatial distributions, they reach maximum values which are clearly lower than the observed IPD and hindcasted HIPOCAS, thus indicating a significant underestimation of the observed precipitation.

The produced relative bias between the observed field and the HIPOCAS, NCEP and ERA fields have shown noticeable differences. The HIPOCAS RB shows most of the study area marked by nearly zero values with some areas of negative values, which indicates a slight underestimation of the observed field. However, the RBs of the NCEP and ERA datasets have shown higher values than the HIPOCAS one over the whole Spanish domain, which shows that the HIPOCAS precipitation field realistically reproduces the observed values, whereas the NCEP and ERA datasets highly underestimate the observed precipitation. The significant improvement in the characterization of the observed precipitation introduced by the HIPOCAS hindcasted in relation to the global reanalyzes is also highlighted by means of the time evolution of the bias and the RMSE averaged over the whole spatial domain. HIPOCAS reproduces the observed IPD field better than NCEP and ERA reanalysis data.

Finally, it is worth noting that the validation over the Iberian Peninsula along with the remarkable improvement relative to global reanalysis data strongly enhance the confidence on the HIPOCAS data. Furthermore, its use can even be helpful in regional climatologic studies which are focused on specific Mediterranean areas which are handicapped by lack of observations, such as offshore areas. Additionally, this chapter confirms that the application of GIS techniques is a useful and promising tool not only for constructing climate maps at different scales, but also for making more exhaustive climatic studies.

13.5. Acknowledgements

This work has been partially supported by the research project CGL2004-01584. The authors wish to thank the following institutions which provided us with data: HIPOCAS data from Ente Público Puertos del Estado (EPPE, Spain), IPD from Spanish Meteorological Service (INM, Spain), NCEP Reanalysis data by the NOAA-CIRES Climate Diagnostics Center, Boulder, Colorado, USA and ERA40 from the European Centre for Weather Medium Forecast (ECWMF).

13.6. Bibliography

[GIB 97] GIBSON J.K., KALLBERG P., UPPALA S., HERNANDEZ A., ÑOMURA A., SERRANO E., ERA description. Technical Report Re-analysis Project Report Series 1, ECMWF, Reading, UK, 1997.

[KAL 96] KALNAY E., KANAMITSU M., KISTLER R., COLLINS D., DEAVEN D., GANDIN L., IREDELL M., SAHA S., WHITE G., WOOLEN J., ZHU Y., CHELLIAH M., EBISUZAKI W., HIGGINS W., JANOWIAK J., MO K.C., KOPELEWSKI C., WANG J., LEETMAA A., REYNOLDS R., JEENE R., JSOSEPH D., "The NCEP/NCAR 40-years Reanalysis, Project", Bull. Amer. Meteorol. Soc., vol. 77, 1996, p. 437-471.

[LUN 04] LUNA M.Y., ALMARZA C., Interpolation of 1961-2002 daily climatic data in Spain, Proceedings of International Meeting on Spatial Interpolation in Climatology and Meteorology. Budapest, Hungary, 2004.

[RAT 03] RATSIMANDRESY A.W., SOTILLO M.G., Reanalisis de 44 años (1958-2001) del clima oceánico y atmosférico en el Mar Mediterráneo: Informe Técnico de la contribución de Puertos del Estado al Proyecto Europeo HIPOCAS. Technical Report. Ente Público Puertos del Estado (EPPE). Madrid, Spain, 2003.

[SHE 01] SHEN S.P., DZIKOWSKI P., LI G., GRIFFITH D., "Interpolation of 1961–97 Daily Temperature and Precipitation Data onto Alberta Polygons of Ecodistrict and Soil Landscapes of Canada", J. Appl. Meteor., vol. 40, 2001, p. 2162-2177.

[SOT 06] SOTILLO M.G., RATSIMANDRESY A.W., CARRETERO, BENTAMY A., VALERO F., GONZÁLEZ-ROUCO J.F., "A high-resolution 44-year atmospheric hindcast for the Mediterranean Basin: Contribution to the regional improvement of global reanalysis", Climate Dynamics, online, DOI 10.1007/s00382-006-0155-3, 2006.

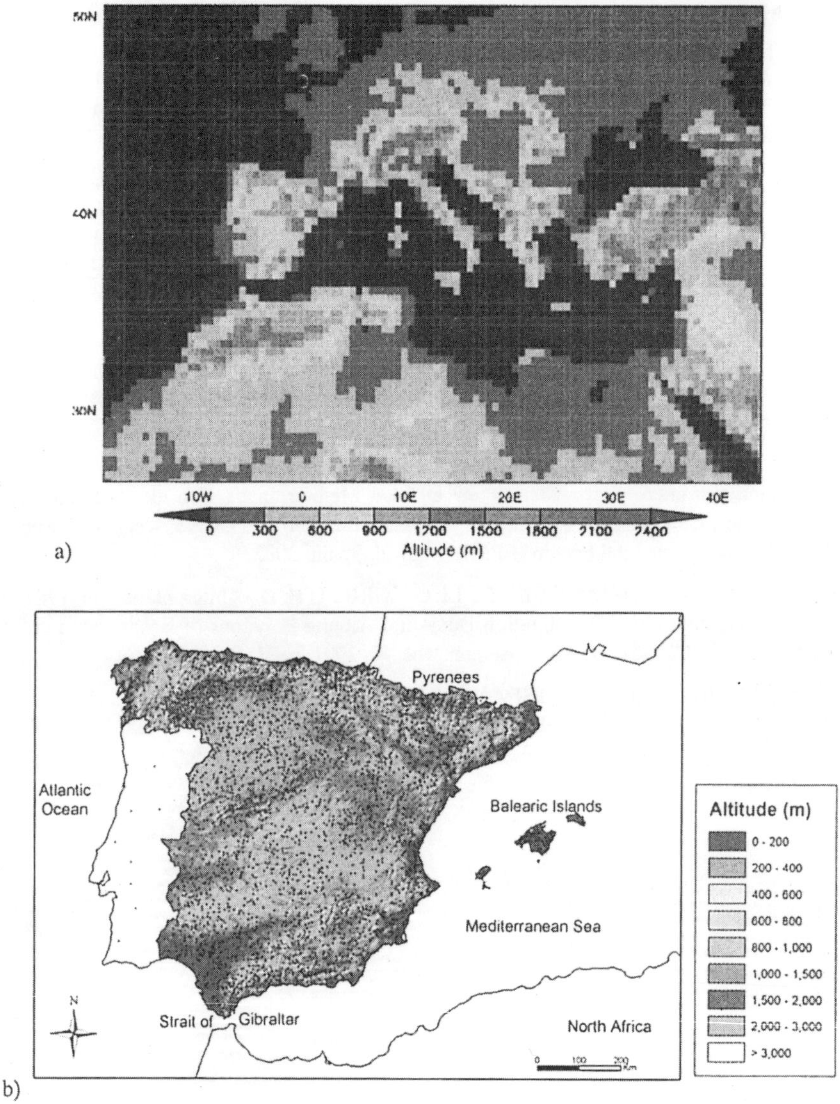

Figure 13.1. *(a) Orography and spatial distribution of the stations over the Iberian Peninsula and the Balearics; (b) Mediterranean HIPOCAS domain and its orography detailed with the HIPOCAS resolution*

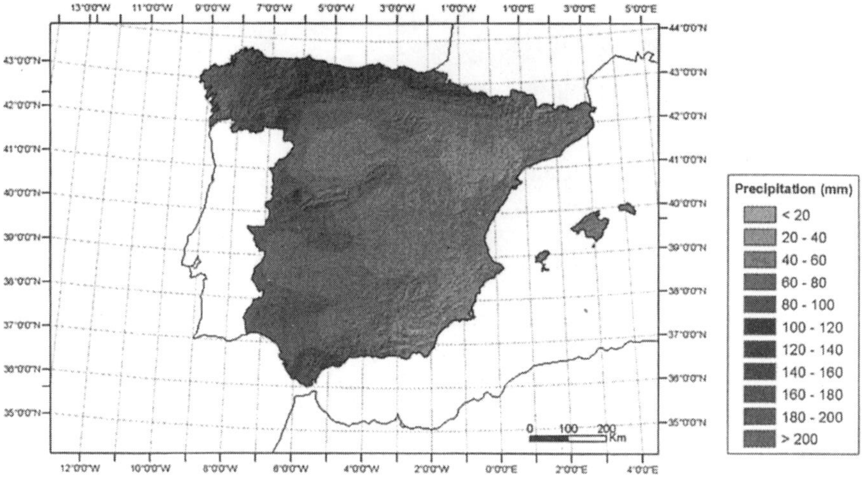
Precipitation (mm)
< 20
20 - 40
40 - 60
60 - 80
80 - 100
100 - 120
120 - 140
140 - 160
160 - 180
180 - 200
> 200
0 100 200 Km

a)

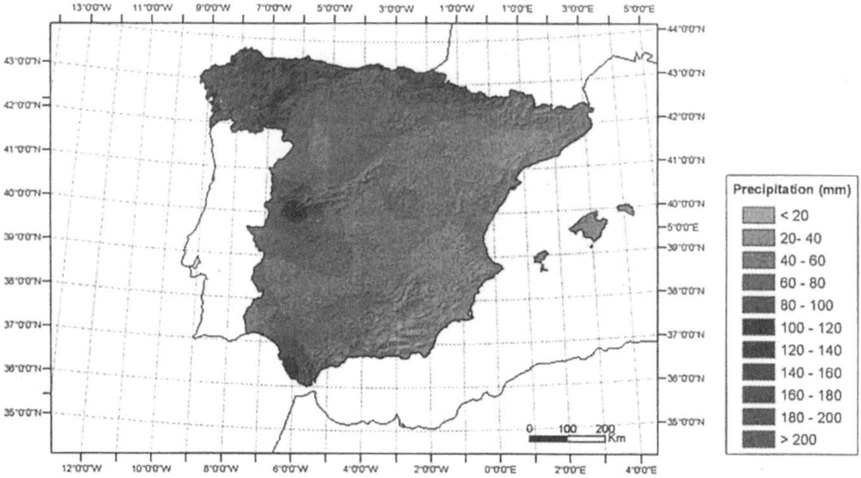
Precipitation (mm)
< 20
20- 40
40 - 60
60 - 80
80 - 100
100 - 120
120 - 140
140 - 160
160 - 180
180 - 200
> 200
0 100 200 Km

b)

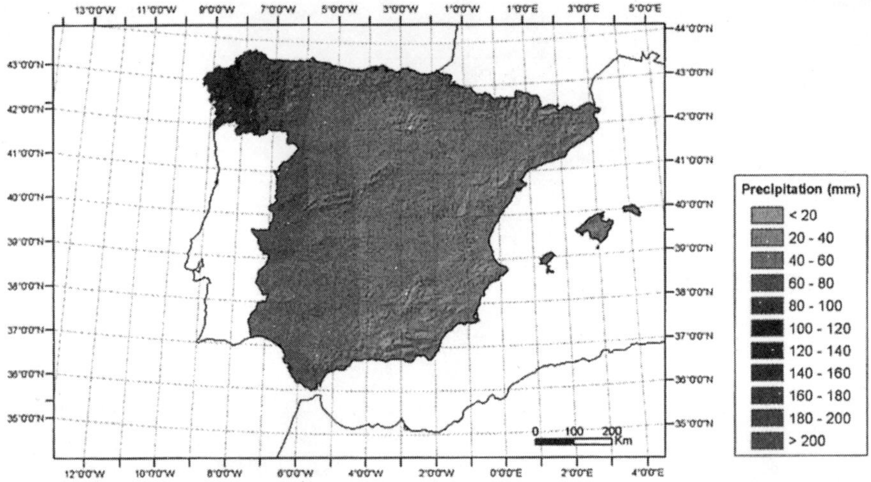

c)

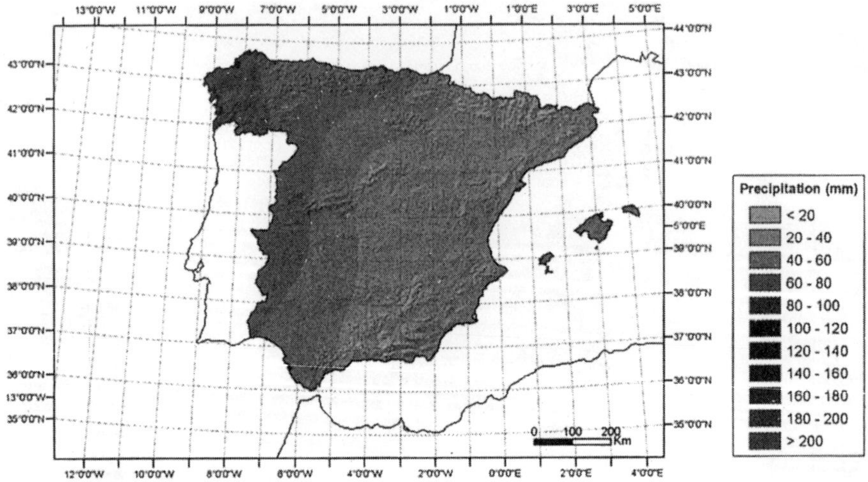

d)

Figure 13.2. *Winter spatial distributions of monthly mean precipitation fields (mm) of: (a) IPD; (b) HIPOCAS; (c) NCEP and (d) ERA*

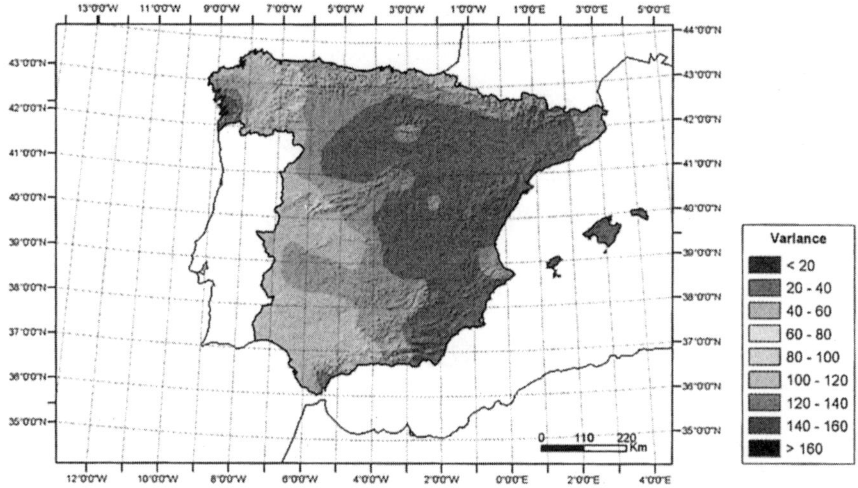
Variance
< 20
20 - 40
40 - 60
60 - 80
80 - 100
100 - 120
120 - 140
140 - 160
> 160
0 110 220 Km

a)

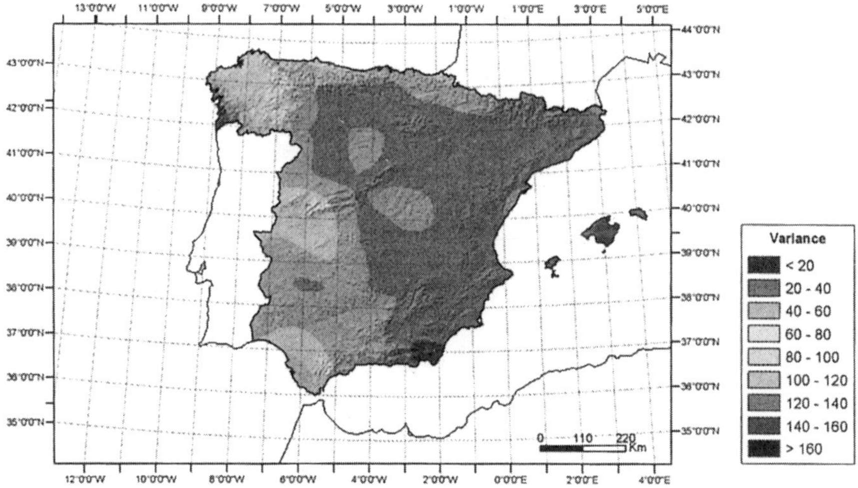
Variance
< 20
20 - 40
40 - 60
60 - 80
80 - 100
100 - 120
120 - 140
140 - 160
> 160
0 110 220 Km

b)

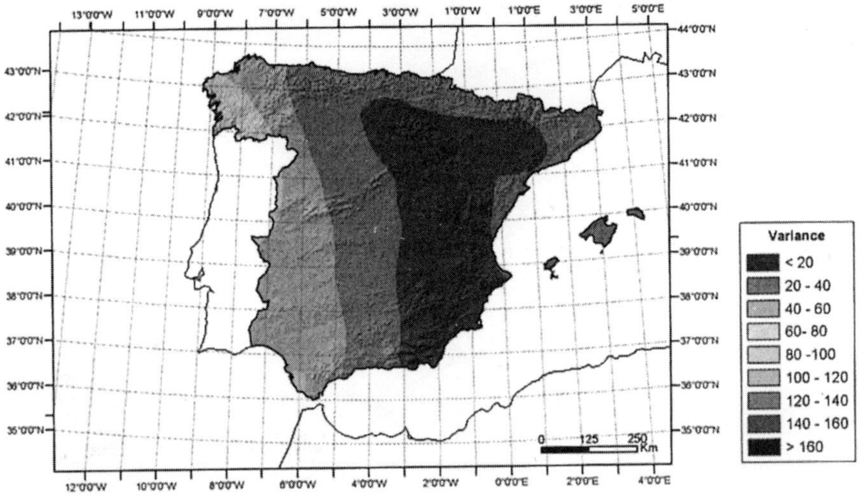

c)

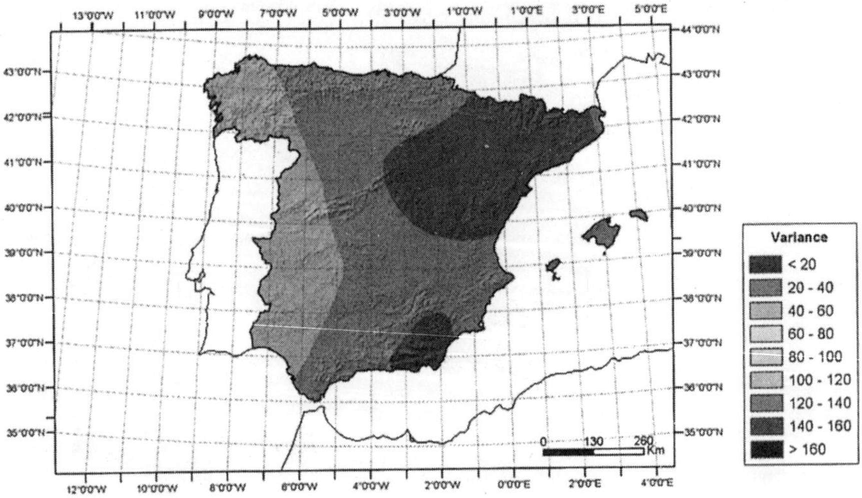

d)

Figure 13.3. *Winter spatial distributions of variance fields (mm^2) of: (a) IPD; (b) HIPOCAS; (c) NCEP and (d) ERA*

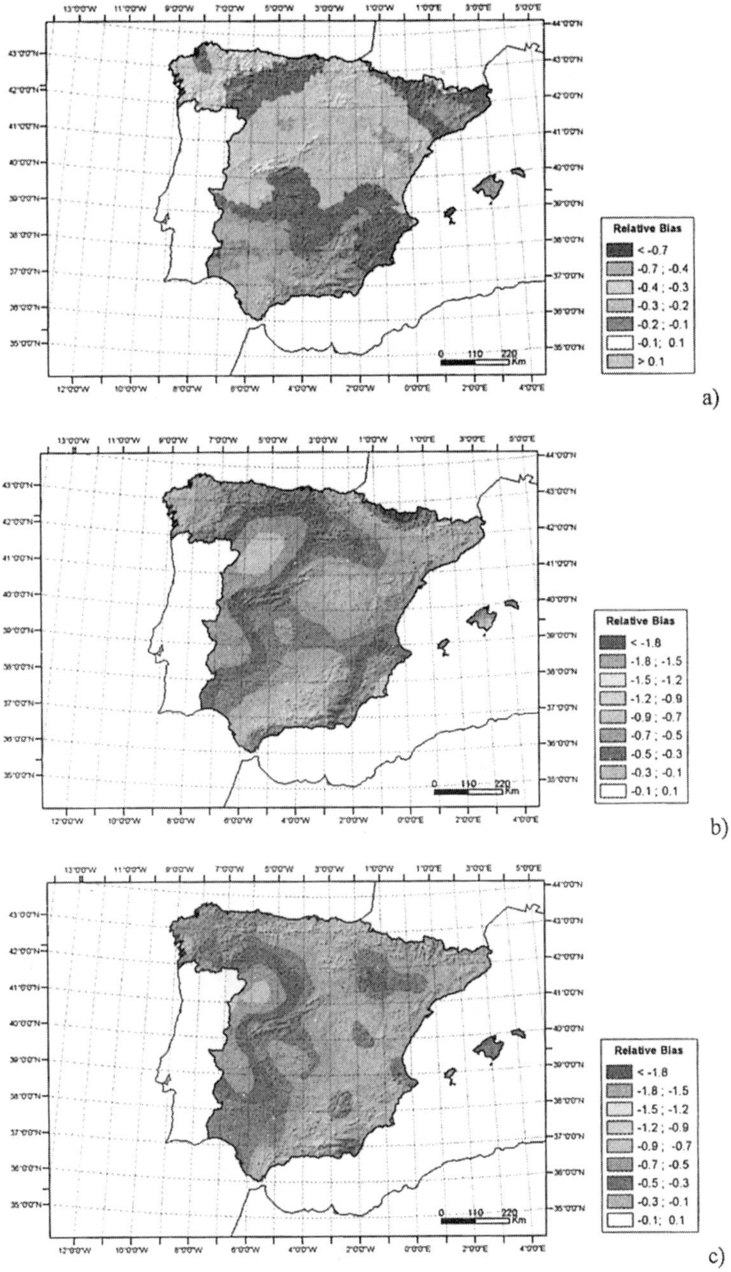

Figure 13.4. *Winter spatial distributions of relative bias (simulated datum minus observed one respect of the observed mean) between: (a) IPD and HIPOCAS fields; (b) IPD and NCEP fields and (c) IPD and ERA fields*

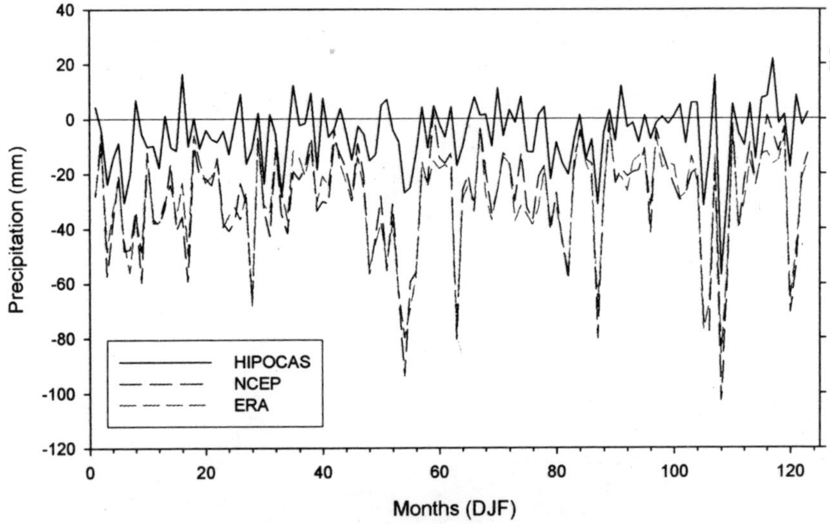

a)

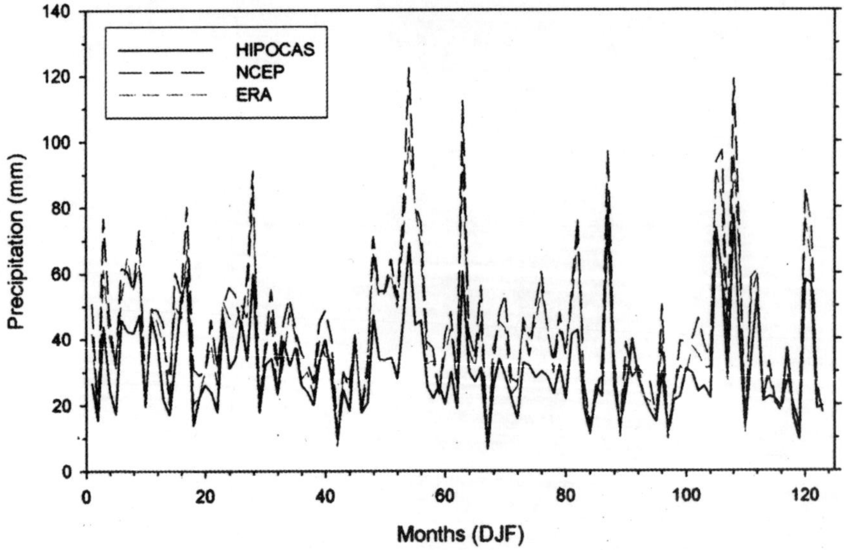

b)

Figure 13.5. *Temporal evolution of: (a) the bias (simulated datum minus observed one) and (b) the root mean squared error averaged over the whole spatial domain for the used datasets*

Chapter 14

Drought Sensitivity Research in Hungary and Influence of Climate Change on Drought Sensitivity

14.1. Introduction

Drought is a very complex phenomena and it does not have a universal definition. Aridity and droughts are natural aspects of the Earth's climate, but aridity is a long-term average feature, whereas droughts are a deviation from the average situation, for a limited period of time. Aridity is defined by long-term low precipitation rates, often accompanied by high evaporation rates, and results in a limited availability of water resources. Drought is a temporary decrease of the average water availability [EUE 06]. Drought is mostly measured by drought indices, which apply the status of the parameters describing the drought situation for the given process. The main problem is that the same degree of drought (in our case the same order of drought indices) could have different influences, different size of damages. Therefore, the vulnerability of the system affected by the drought event could be an even more significant factor than the range and extension of the drought. Vulnerability depends on natural, economic and social factors. A fairly simple method was used to discuss the natural drought vulnerability in our paper. This estimation can serve as a basis to investigate both economic and social vulnerability later.

Chapter written by Sándor SZALAI, Szabolcs BELLA and Ákos NÉMETH.

Vulnerability is the degree to which a system is susceptible to, or unable to cope with, adverse impacts of climate variability and extremes, or climate change. Vulnerability depends on the different features of climate variation to which a system is exposed, its sensitivity and its adaptive capacity [FIN 05]. As a pragmatic approach we assume that vulnerability is the magnitude of change of the output parameter(s) for unit change of the input parameter. The input parameter(s) is (are) the drought index (indices) and the output parameter is the result of the process suffered by drought, for example, yield, number of morbidity, rate of reduced industrial production, etc.

The adaptation is an adjustment in natural or human systems to a new or changing environment. Adaptation to climate change refers to adjustment in natural or human systems in response to actual or expected climatic stimuli and/or their impacts, which moderates harm or exploits beneficial opportunities. Various types of adaptation can be distinguished, including anticipatory and reactive adaptation, private and public adaptation, and autonomous and planned adaptation. The adaptive capacity is the ability of a system to adjust to climate variability and extremes, as well as climate change, to moderate potential damages, to take advantage of opportunities, or to cope with the consequences [FIN 05].

Our main task is to reduce our vulnerability and increase our adaptive capacity. Therefore, first of all we have to determine the present situation. The main problems are the natural disasters in both cases, for example, in the case of Hungary the water caused damages: the drought events, floods and surplus (inland) water.

Probably, the drought causes the largest damages among the natural disasters in Hungary. Drought events were more frequent and lasted longer in the last decades than in any other time before. As a consequence, the value of damages increased significantly.

Therefore, we investigate the vulnerability to the drought events. Most of natural and social processes depend on water. These functions are quite different and therefore drought does not have a generally accepted definition. It is agreed that drought is a phenomenon in connection with relative water scarcity. Many types of drought exist but the most known types are meteorological, hydrological and soil droughts. All drought events begin with meteorological drought, for example, lower than usual quantity of precipitation. The second step in the hydrological cycle is the soil, which is the largest water reservoir in Hungary. The water storage capacity depends on the type of soil and its hydrophysical properties. The additional parameters are the depth of the water table, the surface and subsurface runoff. These last parameters lead to hydrological drought, which is in connection, for example, with the streamflow. The water management of a given area is strongly influenced by land use, too. Practically, it is mostly these four groups of parameters which determine the drought vulnerability of a given region.

14.2. The climate of Hungary

Hungary is situated in Central Europe, within the Carpathian Basin (Figure 14.1). Its climate is mainly determined by the large-scale circulation patterns of maritime, continental and Mediterranean air masses, which are modified by the topography of the basin. The most humid parts of the country are its Western region. Here the precipitation is about twice that of the driest areas of the Hungarian Plain, which is the most important agricultural area of the country. While the highest monthly precipitation values are usually measured in June (60-90 mm), February is the driest month. On the other hand, monthly precipitation can exceed 100 mm or sometimes even 200 mm in any month, but months without any rainfall may occur at any time during the year.

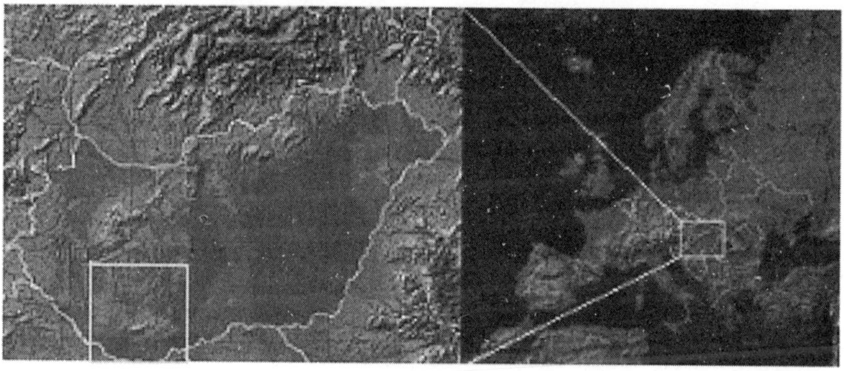

Figure 14.1. *Southern Transdanubium region*

Drought is a general feature of Hungary's climate. The annual mean temperature and precipitation have a great range of variability. [RÉT 98] mentions the damages caused by droughts, floods and inland waters according to memories of contemporary chronicles, letters, clerical sources, etc.

Drought was more frequent in the 1990s than anytime before. This led to the great increase of damages. The longest was between 1983 and 1995. The severe drought and the deficit of water became continuous and the annual amount and distribution of precipitation was normal only during 3 years (1987, 1988 and 1991) out of 13. In the first years of the 21st century there were also droughts. An extremely severe drought was all over of Europe in 2003. The period between May and August was the hottest of the previous 100 years. This accentuates the consequences of lack of precipitation. Summer heat (a daily maximum temperature of above 30°C) was recorded on 45 days (country average). Crop yields decreased by more than 30-40% and estimated losses exceeded 10 billion HUF.

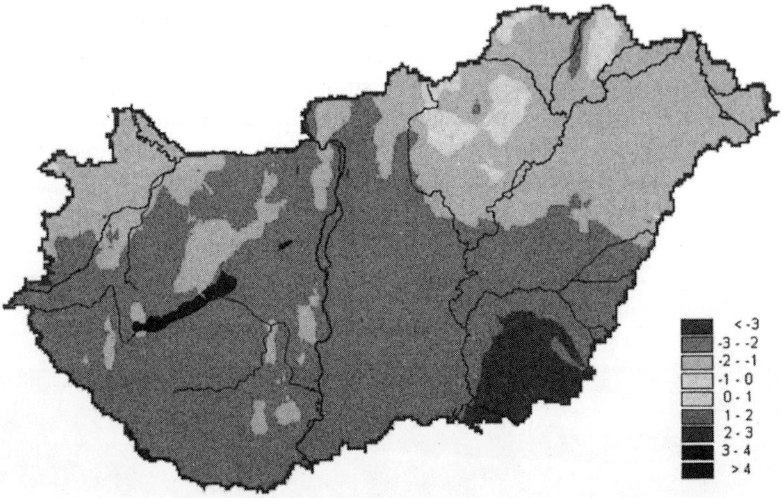

Figure 14.2. *6-month SPI index for Hungary in August 2003*

14.3. Method

The original method of classification was used by [WIL 02]. This method was modified for the Hungarian conditions and [BEL 2005] have set up a method for measuring drought vulnerability. They considered Somogy county as a sample area. Considering the results, an improved drought vulnerability map of Southern Transdanubium was constructed. We provide a short description below.

14.3.1. *Soil parameters*

Soil data was obtained from the AGROTOPO database of the Research Institute of Soil Science and Agrochemistry of the Hungarian Academy of Science. The hardest part of the examination was the determination of the weight factors of soil data. We used data regarding soil texture, organic content, topsoil thickness, mineral content, type of rocks and water management.

14.3.1.1. *Mineral content*

Minerals play an important role for the soil and the rocks. They determine the number and proportion of chemical elements, the physical and chemical behavior of the substance, and the impact on living organisms. Land classification is a function of clay minerals. The clay minerals are the most important class of the mineral parts of the soil. The nutriments and water management of the soil depend on the clay minerals quality and quantity.

14.3.1.2. *Soil texture*

Soil texture is the integrated determinant of the scale distribution of the organic and mineral components of the soil. It indicates the quantity and distribution of different particle size of mechanical content. Similarly to the mechanical content the soil texture is very stable in time. It determines the physical features, affects the mechanical features of the soil (plasticity), absorption capability, chemical and physical features.

14.3.1.3. *Water management*

Water management was considered according to the sinkhole capacity, permeability and water storage capacity.

The sinkhole capacity is indicated by the amount of water that seeps into the soil per unit of time. Since the infiltration is made up of two parts (pores full of air filled up with water + water conduction towards deeper soil layers) the sinkhole capacity is always bigger – especially in the case of dry soil – than the water leading capacity [STE 81].

The permeability is measured by the amount of water that passes through the cross section of the saturated soil per unit of time. The permeability of the whole section is determined by the value of the worst layer. The distance of this layer from the surface is very important. If the bad water conducting layer is just 4-5 cm under the surface, it can swell the water to the upper layers during heavy rains, which can cause surface runoff. It can be said that the upper layers of the soil (10-20 cm) – i.e. the ploughed layer – have the crucial role in water leading capacity [GÓC 69].

The water storage capacity of the soil allows plants to satisfy their continuous water demand from the very unsystematical (very often its temporal and spatial distribution is irregular) precipitation activity during their growth periods. The same meteorological situation leads to different ecological outcomes (in crops and yield) on different soils, as the record of the most recent past shows. It is understandable that years of drought overvalue the importance of the soil and this shows that the soil is the biggest natural water storage in Hungary.

14.3.1.4. *Topsoil thickness*

This layer is important for plants and agriculture because they get the water and the necessary nutrients from it.

14.3.1.5. *Organic content*

The organic content plays a very important role (considering both quality and quantity) in the development of soil fertility. The organic content of the soil – humus – is not a uniform substratum: it is a mixture of many organic components with different chemical contents and physical behavior. We classified it according to the total amounts per hectare.

14.3.1.6. *Type of rock*

Rocks give basic sources for soil formation. Their physical properties, chemical and mineral content affects the soil parameters.

14.3.2. *Precipitation*

Precipitation has a very important role – with respect to climate – in drought development. Therefore, we first processed the precipitation data. We used long time series of 53 rain gauge stations, which are in Somogy, Baranya and Tolna County belonging to the Hungarian Meteorological Service. The precipitation data was homogenized with MASH [SZE 99].

For the determination of drought vulnerability, we used the spatial averages of the long-term data in the given territory. We chose a period that was long enough for climatological research and in which there was no change in the measuring network. We made precipitation maps to three such time intervals, namely: 1951-1980, 1961-1990 and 1971-2000. We made the interpolation with the kriging method. According to the three 30-year periods the big annual amounts of precipitation (> 700 mm) decreased substantially in the SW part of the region.

14.3.3. *Groundwater*

The groundwater data was obtained from the archives of hydrological data of VITUKI (Water Resources Research Centre) and from hydrographical yearbooks. We examined the following periods: 1951-1980, 1961-1990, 1971-2000. Since 1971 much more stations have measured groundwater than between 1951-1971 and the interpolation was kriging too. The decrease of groundwater is demonstrable; the most significant occurrence is in Belső-Somogy, whereas the opposite process can be observed in Nagyberek.

14.3.4. *Land use*

Land use data was obtained from 1:100.000 CORINE (1994) Land Cover database, of the Institution of Survey and Remote Sensing (FÖMI). The land use categories [COR 94] were simplified significantly (five categories).

14.3.5. *Slope and aspect*

We used the Digital Terrain Model made by the Cartographer Service of the Hungarian Defense Ministry (DDM-100) to represent the surface elevation information of the territory. With respect to the aspect we took into account the four main points of the compass and plain lands were treated separately. The aspect's categories are those used in physical geography.

Figure 14.3 represents the flowchart of determination of drought vulnerability. We made our maps with a similar method in our previous works [BEL 05] but then we used an extra weighting and reclassifying method. We will see later that there is no significant difference between these results.

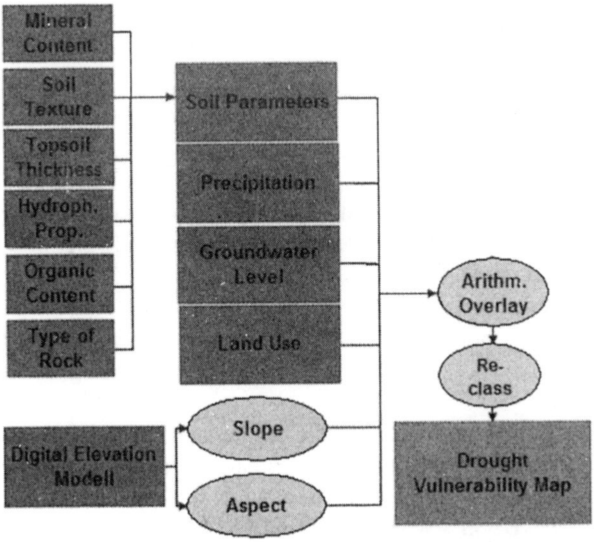

Figure 14.3. *Flowchart of determination of drought vulnerability*

We supposed in the examination that, except for precipitation and groundwater data, other parameters are constants. This is acceptable for a shorter period of time, but if we would like to examine a longer period, change of land use must be considered too and there may be changes in the soils too. Considering these is not easy because the old maps are available only in papers based, analog format. If we wanted to use them, we would have to perform a difficult digitalization process. We examined the impact on vulnerability of a change of the two natural variables.

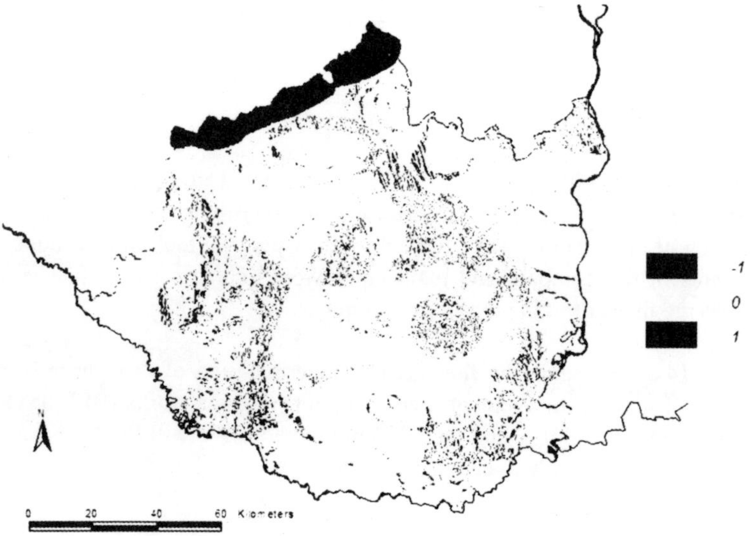

Figure 14.4. *Changes of drought vulnerability according to the distinction of maps of 1951-1980 and 1961-1990*

Figures 14.4 and 14.5 are tendency maps. We made the region's drought vulnerability map according to the precipitation and groundwater data of 1951-1980. The same method was used for 1961-1990 and 1971-2000. After this we "subtracted" the other two maps from the map of 1951-1980, which were made according to the other two 30-year averages. In this way we can see the changes of vulnerability.

The situation turned more unfavorable on the lands of Belső-Somogy, Baranyai-dombság, the Southern parts of Lake Balaton, Southern Mezőföld and other territories of the region (Figure 14.4). These lands were placed into worse categories of the 5-class scale. Besides worse lands (marked by red color) there are better ones (marked by blue), but they are much smaller.

If we take the aforementioned subsection with data from 1971-2000, we obtain a similar picture (Figure 14.5). Both unfavorable and favorable lands occur more expressively. For example, the lands of Nagyberek belong to the latter category. As we mentioned, we slightly changed the calculation method of drought vulnerability. We carried out the examination with this too and we obtained a similar picture to that of Figures 14.4 and 14.5. The number of lands in the two categories (favorable and unfavorable) is smaller.

The question is which method is closest to reality? Verification would help answering it, but the difficulty in accessing data hinders the investigation. However, this is already in progress with the received data.

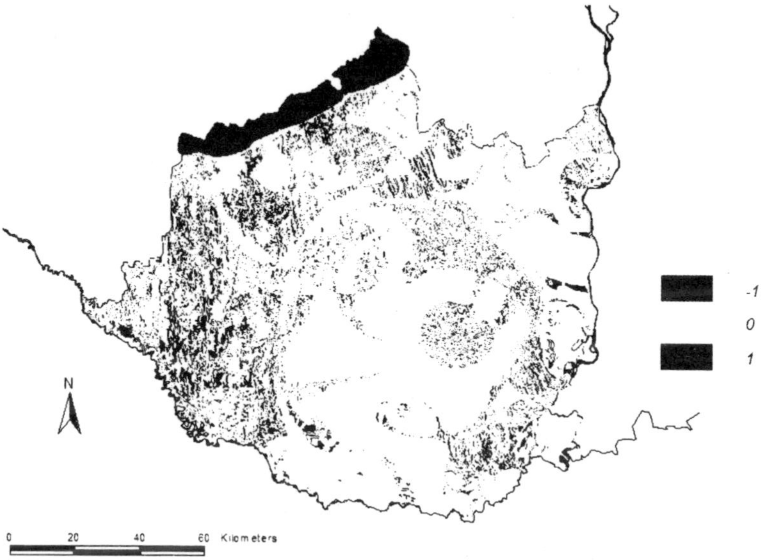

Figure 14.5. *Changes of drought vulnerability according to the distinction of maps of 1951-1980 and 1971-2000*

14.4. Conclusion

Drought influences many different segments of life; indeed, if we consider only agriculture, we can state that almost each plant is vulnerable to drought since horticultural vegetables and plants need irrigation. In the protected areas the lack of water, and drought is a great problem. The drying of wetlands can cause huge damages. We showed that climatical tendencies at the moment indicate the growth of drought vulnerability. Our method contributes to an easy determination of lands that are sensitive to change of precipitation and groundwater levels. This can be of great help for farmers when making their decision regarding plants that they are going to cultivate in these areas.

14.5. Acknowledgements

Special thanks to Péter Fákó (American Appraisal Associates, Hungary) for his help in proofreading.

14.6. Bibliography

[BEL 05] BELLA S.Z., *et al.*, 2005: "Examination of Drought Vulnerability with GIS Tools: Somogy County Case Study", Remote Sensing & GIS for Environmental Studies, p. 209-217.

[COR 94] CORINE, 1994: Land Cover Technical Guide. – European Commission, Luxembourg.

[EUE 06] EU Environment Council, Information Note, 27 June 2006.

[FIN 05] Finnish National Strategy for Adaptation to Climate Change, 2005.

[GÓC 69] GÓCZÁN L., KAZÓ B., 1969: "A mérnökgeológiai-vízgazdálkodási térképezés új módszerei és felhasználási területei". Földrajzi Értesítő 18, pp. 409-417.

[RÉT 98] RÉTHLY A., 1998: "Időjárási események és elemi csapások Magyarországon 1801-1900 I., II. Kötet", OMSZ, Budapest.

[STE 81] STEFANOVITS P., 1981: "*Talajtan*", Mezőgazdasági kiadó, Budapest.

[SZE 99] SZENTIMREY T., 1999: "Multiple Analysis of Series for Homogenization (MASH)", Proceedings of the Second Seminar for Homogenization of Surface Climatological Data, Budapest, Hungary; WMO, WCDMP-No. 41, p. 27-46.

[VÁR 00] VÁRALLYAY G.Y., 2000: "The role of soil and soil management in drought mitigation" in Proceedings of the Central and Eastern European Workshop on Drought Mitigation, edited by Vermes László and Szemessy Ágnes, 2000, Budapest, RePRINTSTudio Ltd.

[WIL 02] WILHELMI O.V., WILHITE D.A., 2002: "Assessing Vulnerability to Agricultural Drought: A Nebraska Case Study", 2000 — Natural Hazards, vol. 25, p. 37-58.

[WOR 02] WORLD METEOROLOGICAL ORGANIZATION, 2002, "Reducing vulnerability to weather and climate extremes", Geneva.

Chapter 15

First Steps Towards a New Temperature Climatology of the Greater Alpine Region (GAR)

15.1. Introduction

The Greater Alpine Region (GAR) covering the area between 4-19°E and 43-50°N and an altitude range between 0 and more than 4,000 m.a.s.l. offers a challenging climate which is worth studying in any detail. However, it is surprising that up to now there has been no comprehensive Alpine temperature climatology covering the whole region. This is why the ECSN project HRT/GAR aims to produce gridded monthly temperature maps for this region in a spatial resolution of at least 1 km^2 (or 1 min). A second attribute for each grid-cell, i.e. the local vertical lapse rate, will enable corrections of still existing sub-grid effects due to steep orography. The period under investigation will be 1961-1990. This chapter describes the first steps of the project.

The GAR studied comprises 12 countries which are: France, Switzerland, Liechtenstein, Italy, Germany, Czech Republic, Slovakia, Hungary, Austria, Slovenia, Croatia and Bosnia-Herzegovina. The number of data holders (national and/or regional meteorological and hydrographical services, etc.) is somewhat higher. For precipitation a supranational Alpine Climatology covering the majority of GAR over the period 1971-1990 was prepared by Frei and Schär (1998) and Schwarb (2001). For air temperature, however, a similar investigation is still

Chapter written by Wolfgang SCHÖNER, Ingeborg AUER and Reinhard BÖHM.

missing, although a number of national and/or regional investigations exists (e.g. Hidrometeorološki zavod Republike Slovenije, 1995, Auer *et al.*, 2001, Gajić-Čapka *et al.*, 2003).

15.2. Data

Monthly air temperature normals (period 1961-1990) of 1,723 climate station from all Alpine countries as well as from countries which are quite close to the Alps are the data source for this study. Additionally, relevant metadata (longitude, latitude, altitude, algorithms for the calculation of mean temperature) were provided by the 20 data holders. For most of the stations the calculated normals were based on series with full temporal coverage (30-year observations). However, to increase the spatial density, additional stations with at least 20 years were included (>=15y for high elevations), adjusted to 1961-1990 by using only climatologically meaningful nearby sites. Some metadata information of station network is shown in Table 15.1. The spatial distribution of climate stations is shown in Figure 15.1. It is obvious from that figure that spatial coverage varies remarkably between the different countries, despite the large effort put in the data sampling and data preparation.

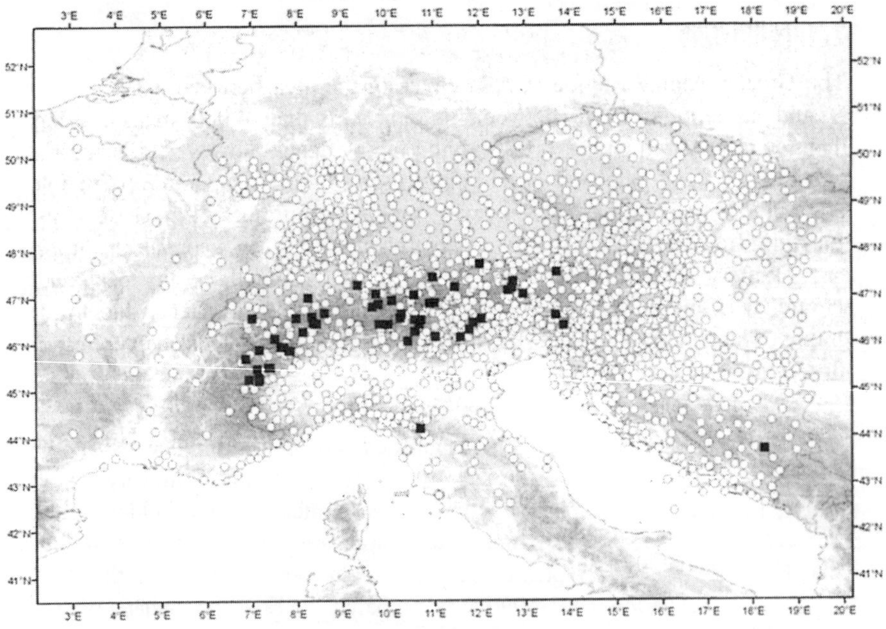

Figure 15.1. *Map of the GAR studied showing climate stations used in the study (black boxes are stations higher than 1,800 masl)*

n	nat.	data producer	provided by: organisation	means-calc	time
227	AT	ZAMG	ZAMG	1971-1990: (7+14+21+21)/4 1961-1970: (7+19+tm+tx)/4 both pre-adjusted to TRMs	MLT
171	AT	CHS	HZB	(7+14+21+21)/4 pre-adjusted to TRMs	MLT
29	BA	METEOBIH	METEOBIH	(7+14+21+21)/4	MLT
16	BA	METEOBIH	METEOBIH	(7+14+21+21)/4	MLT
91	CH	MeteoSwiss	MeteoSwiss	pre-adjusted to TRMs	CET+30'
121	CZ	CHMI	CHMI	(7+14+21+21)/4	MLT
304	DE	DWD	DWD	1961-1986: (7+14+21+21)/4 1987-1990: (7.30+14.30+2*21.30)/4	MLT CET
45	FR	MF	MF	(tm+tx)/2	
152	HR	DHMZ	DHMZ	(7+14+21+21)/4	MLT
31	HU	OMSZ	OMSZ	(7+14+21+21)/4	MLT
141	IT	?	UNIPAD	(tm+tx)/2	
140	IT	?	ISAC-UNIMIL	(tm+tx)/2	
36	IT	?	SMI	(tm+tx)/2	
41	IT	?	UNIPAVTOR	(tm+tx)/2	
9	IT	?		(tm+tx)/2	
14	IT	?	UTOR-CD	(tm+tx)/2	
108	SI	ARSO	ARSO	(7+14+21+21)/4	MLT
22	SK	SHMU	SHMU	(7+14+21+21)/4	MLT

Table 15.1. *Metadata of climate station network used in the study (AT: Austria, BA: Bosnia-Herzegovina, CH: Switzerland, CZ: Czech Republic, DE: Germany, FR: France, HR: Croatia, HU: Hungary, IT: Italy, SI: Slovenia, SK: Slovakia, tm: minimum temperature, tx: maximum temperature, MLT: mean local time, CET: Central European time)*

All data was carefully checked for data quality. Only single cases required the correction of errors with respect to double station information (two different values for what is in fact one station), mixtures of temperature maximum and mean as well as minimum and mean, outliers and precise geographical information. A greater number of errors occurred concerning metadata. Especially the accuracy of geographical coordinates (longitude and latitude) turned out to be in adequate for the planned final validation of spatial modeling. 98 out of 100 cases of lat-long-errors resulted in very large differences between the altitudes of stations and the elevation model.

Air temperature normals were adjusted to the common mean calculation (7+14+21+21)/4. Several different algorithms for mean calculation are in use within the GAR for the period 1961-1990 such as (7+14+2*21)/4, (7+19+Max+Min)/4, (Max+Min)/2, (7.30 + 13.30 + 2x21.30)/4. The algorithms used by the individual data holders of GAR can be seen in Table 15.1. Figure 15.2 shows that the differences between different algorithms of mean computation are quite remarkable, especially the (Max+Min)/2 formula, which is quite commonly used, could result in rather large differences from the 24 hours mean.

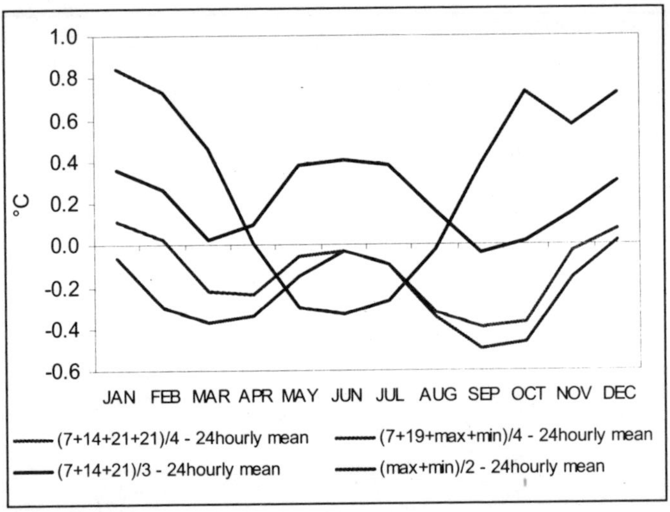

Figure 15.2. *Comparison of the mean monthly temperature obtained by using different algorithms to calculate the monthly mean at the Austrian station of Puchberg. All values are shown as differences from the "true" (24 hours) mean*

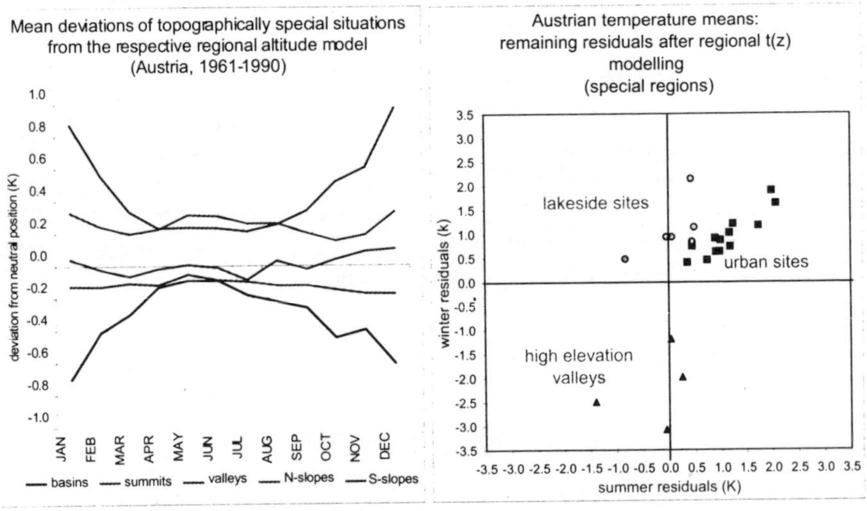

Figure 15.3. *Air temperature specifics due to different topography and land use derived from climate stations in Austria*

In a special study the climate stations from Austria were analyzed for site specifics of air temperature with respect to exposition and land use of station surroundings. Some clear features appear from Figure 15.3. Most remarkably, the southern slope stations are about 1.5°C warmer compared to basin stations in winter, whereas the difference in summer is only 0.4°C. Moreover, the northern slope stations are warmer than those in the South during summer. This finding (which on first impression seems to be surprisingly) can be explained by the difference in potential global radiation in summer between the southern and northern slopes of the Alps. For land use the specifics most pronounced differences can be seen for urban stations with higher temperature during winter and summer. Climate stations at lake sites are characterized by increased temperatures during winter.

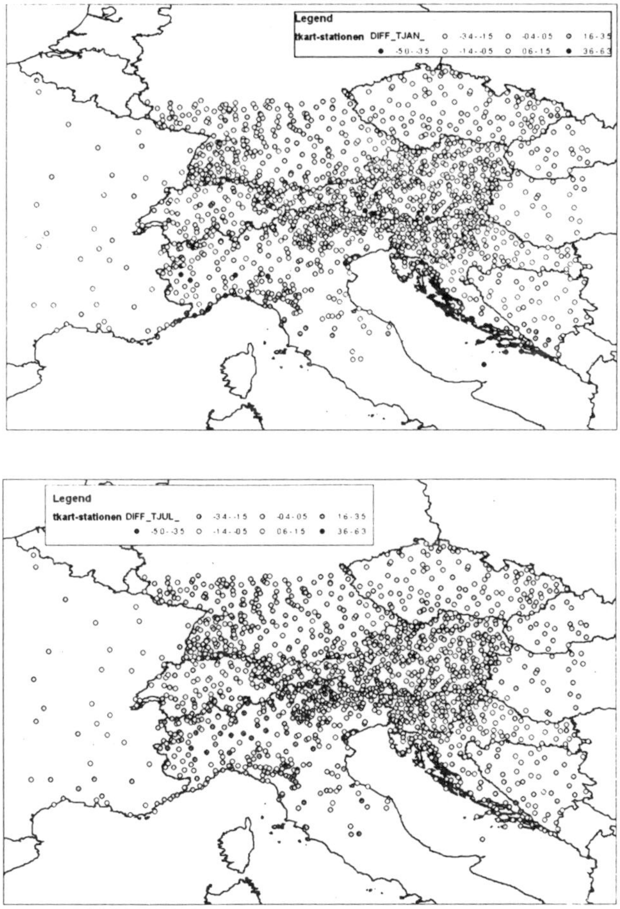

Figure 15.4. *Residuals of linear regression between air temperature and altitude for January (top) and July (bottom) for GAR*

15.3. Spatialization

Several methods have been described for climate spatialization by national approaches. [TVE 00] gave an overview of different methods for spatial modeling of climate elements. For the majority of investigations the methods which are used split the modeled variable into a deterministic as well as a stochastic part (e.g. residual kriging; see [TVE 98, TVE 00]. Such a method is especially powerful for air temperature because of its strong dependency on altitude (the deterministic part). Additional useful predictors of air temperature are, for example, latitude, longitude, distance to coast or weather types (see [TVE 00]). More sophisticated deterministic methods use, for example, coding of the orography as an additional predictor. Such coding of the orography can be done by means of principle component analysis of the DEM (digital elevation model) as implemented, for example, in the AURELHY method [BEN 86].

For the GAR the best spatialization method has to be determined by detailed validation procedures. In this chapter we started with the modeling of the deterministic part by linear regression between air temperature and altitude for the entire region of GAR. Residuals of the linear regression at the climate station sites (Figure 15.4) clearly show a spatially clustered pattern throughout the year (for simplicity Figure 15.4 only shows the months of January and July which cover the annual range sufficiently). This finding supports a regionalized deterministic modeling of monthly air temperatures by applying linear regressions for sub-regions. The effect of regionalization on linear regression between air temperature and altitude clearly appears from Figure 15.5 too. Air temperature variance remarkably decreases for individual sub-regions (derived from the residual map in Figure 15.4) with respect to the entire GAR. If, however, the deterministic modeling of air temperature is done by a multiple linear regression using altitude, longitude and latitude as predictors the picture changes (Figure 15.6). Whereas for January some better model performance can be obtained with respect to maximum residuals (but not for the RSME and R^2), for July the results for some regions are worse for maximum residuals, for region MM and even for the RMSE. It appears from that result that the sub-regionalization is partly captured by the introduction of latitude and longitude as additional predictors of air temperature.

15.4. Summary and outlook

This chapter clearly shows the importance of a spatialization pre-phase of data preparation and data quality control for spatial modeling approaches. The detailed metadata information of climate stations can significantly improve the spatialization results if, for example, topographical as well as land use specifics can be derived. With such information topographical as well as land use specifics can be much better introduced into the model. The first steps of spatial modeling show that altitude, longitude as well as latitude are powerful predictors of the Alpine air temperature field. However, further modeling attempts are necessary within the ECSN GAR-HRT project in order to fall below the desired model errors. In a further step, the GAR air temperature climatology 1961-1990 will be used for calculating the monthly absolute air temperature fields back to 1760. This will be achieved by matching the climatology field with the air temperature anomaly fields of the HISTALP database (see [AUE 01 for details on HISTALP).

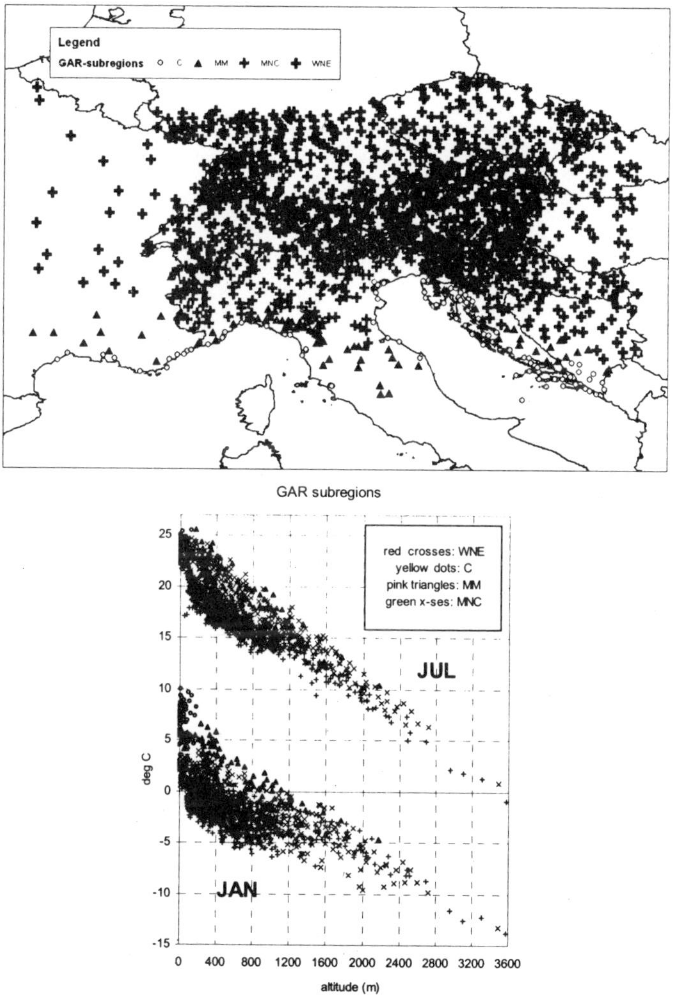

Figure 15.5. *Relation between air temperature and altitude for individual sub-regions (bottom). The sub-regions (top) were derived from the residual maps shown in Figure 15.4*

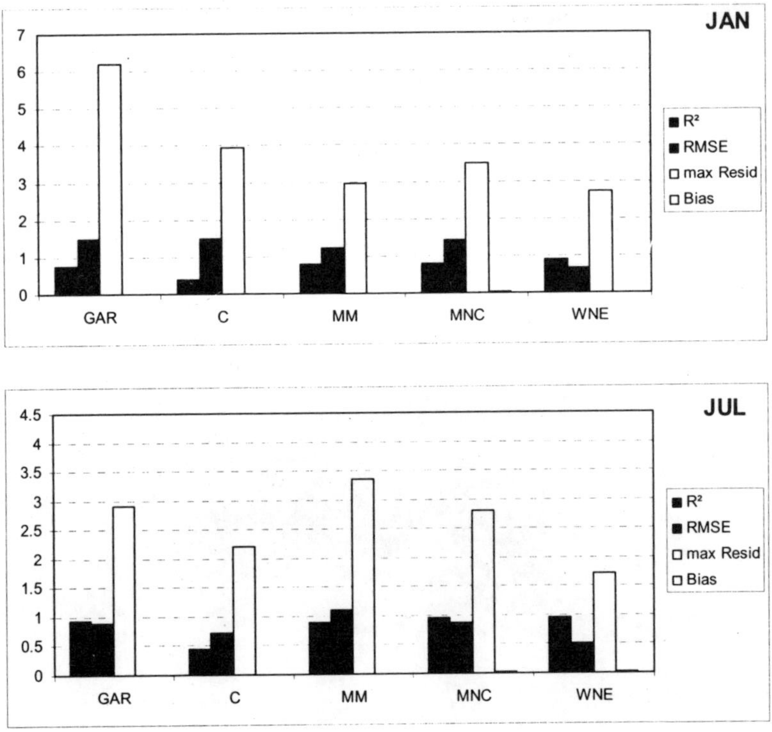

Figure 15.6. *Statistical measures of model performance for the deterministic modeling of air temperature from altitude, latitude and longitude. The bars are for the entire GAR (left) as well as for the individual sub-regions as shown in the map of Figure 15.5*

15.5. Acknowledgements

We would like to thank G. Müller-Westermeier, V. Kveton, T. Cegnar, M. Dolinar, M. Gajić-Čapka, K. Zaninović, M. Maugeri, M. Brunetti, T. Nanni, M. Carrer, L. Mercalli, Z. Majstorovic, M. Begert, J.-M. Moisselin, J.-P. Ceron, O. Bochnicek, Z. Bihari and P. Nola for providing us the data, background knowledge and metadata information. The work was financially supported by the project ALP-IMP which is supported by European Commission (EVK2-CT-2002-00148) and the ECSN project "HRT-GAR: High Resolution Temperature Climatology in Complex terrain – demonstrated in the test area Greater Alpine Region (GAR)".

15.6. Bibliography

[AUE 01] AUER I., BÖHM R., MOHNL H., POTZMANN R., SCHÖNER W., SKOMOROWSKI P., 2001: Öklim – Digitaler Klimaatlas Österreichs, CD Zentralanstalt für Meteorologie und Geodynamik.

[BEN 86] BENICHOU P., 1986. Cartography of statistical pluviometric fields with automatic allowance for topography. 3rd Int. Conf. on Stat. Climatology, Vienna, Austria.

[FRE 98] FREI C., SCHÄR C., 1998: A precipitation climatology of the Alps from high-resolution rain-gauge observations. Int. J. Climatol., 18, 873-900.

[GAJ 03] GAJIĆ-ČAPKA M., TADIC M.P., PATARCIC M., 2003: Digitalna godišnja oborinska karta hrvatske. Hrvatski meteorolološki časopis 38, 21-33.

[HID 95] HIDROMETEOROLOŠKI ZAVOD REPUBLIKE SLOVENIJE, 1995. Klimatografija Slovenije 1961-1990. Temperatura zraka.

[SCH 01] SCHWARB M., 2001: The Alpine Precipitation Climate. Evaluation of a high Resolution Analysis Scheme Using Comprehensive Rain gauge data. Diss. ETHZ 13'911. Zürcher Klimaschriften, Heft 80, Institut für Klimaforschung ETH, Verlag Institut für Klimaforschung ETH Zurich.

[TVE 98] TVEITO O.E., FOERLAND E., 1998: Spatial interpolation of temperatures in Norway applying a geostatistical model and GIS. DNMI, Report No. 26/98.

[TVE 00] TVEITO O.E., FOERLAND E., HEINO R., HANSSEN-BAUER I., ALEXANDERSSON H., DAHLSTRÖM B., DREBS A., KERN-HANSEN C., JONSSON T., VAARBY LAURSEN E., WESTMAN Y., 2000: Nordic temperature maps. DNMI, Report No. 09/00.

[TVE 02] TVEITO O.E. and SCHÖNER W., 2002. Applications of spatial interpolation of climatological and meteorological elements by the use of geographical information systems (GIS). COST719 – The use of geographical information systems in climatology and meteorology. Report No. 1/WG2 Spatialisation. ISSN 0805-9918, Norwegian Meteorological Institute, 2002.

Chapter 16

XRWIS: A New GIS Paradigm for Winter Road Maintenance

16.1. Introduction

The current paradigm in the UK (and most other countries with cold winter climates) for the prediction of ice and snow on roads was developed in the 1980s by the University of Birmingham and the UK Met Office for the benefit of highway engineers [THO 91]. Currently, in the UK about £150m is spent in an average winter to spread salt in order to prevent ice formation and remove snow from our major roads. Around the globe approximately $10 billion are spent every winter to keep roads open and safe. Without the current Road Weather Information Systems (RWIS) the cost of winter road maintenance would be much higher and there would be many more accidents and delays on slippery roads in winter. Today there are more than 200 Highway Authorities (HAs) responsible for the winter maintenance of roads in the UK. More than 1,000 road weather outstations have been installed and virtually all salting routes have been thermally mapped [THO 83]. In total there are more than 3,500 Salting Routes in the UK and approximately 120,000 km (30% of the 400,000 km) of main roads are salted. Each cold night in winter we spend about £3m to keep the nation's roads open and we spend about £1,250/km each winter to keep our main roads open and safe. Thornes [THO 00] has shown that for every £1 spent on winter maintenance about £8 are saved and hence we could think that the current paradigm is perfectly adequate, but the methodology behind the current RWIS is 20 years out of date because technology has moved on and we can

---

Chapter written by John E. THORNES, Lee CHAPMAN and Steve WHITE.

now reduce costs and improve performance by updating our current systems to the next generation road weather information system: XRWIS.

16.2. The current RWIS paradigm in the UK

If salt can be spread on our roads before ice forms then much less salt is required (presalting at 10 g/m^2) than if the ice has to be melted by the salt (de-icing at 40 g/m^2). Similarly, if salt is spread in advance of a snowfall (usually at 40 g/m^2) then small amounts of snow will be melted quickly by the salt and greater amounts of snow can more easily be removed as the snow will have been prevented from freezing onto the road, hence, the need for accurate weather prediction so that winter maintenance activities can be proactive rather than reactive.

In the early 1980s, before the widespread use of RWIS, the Met Office used to issue Road Danger Warnings to Highway Authorities [THO 77]. These would be simple statements such as:

Road surface temperatures are expected to fall below zero at 2 am and ice is expected to form on most of the roads in the region.

These warnings were usually sent by fax to HAs and were effectively issued at cost.

In the early 1980s "Ice Detection Systems" were introduced to the market. These comprised a number of road weather "outstations" which could be interrogated by an "instation" computer at a HA central depot. However, these systems could only detect when ice had formed on the road sensor by which time it was too late to presalt the roads. Also, the output from these sensors could be misleading as salt is hygroscopic. For example, an ice detection system could give a "critical" warning when the road surface temperature falls below zero and the road is "wet". If the road is "wet" due to residual salt going into solution as the air humidity rises above 80% then this "false alarm" could lead to the road being salted again unnecessarily!

Another problem was that the road sensor was only able to give the road surface condition and temperature at one location within a few centimeters and it was not known how representative this was of the rest of the road network. The concept of thermal mapping was developed in the UK at the University of Birmingham to give a spatial estimate of the likely minimum road surface temperature along a route. The first commercial thermal map was produced for the A9 in the Highland Region in 1984. Also, at this time the concept of climatic zones within a HA was developed based on thermal map data.

In the remainder of the 1980s a flurry of activity led to the installation of what are called "Ice Prediction Systems" in HAs. The concept was an improvement over ice detection for a number of reasons. Firstly, the Met Office in conjunction with Thermal Mapping International (TMI was a spin-out company from the University of Birmingham) developed a new forecast product to replace Road Danger Warnings called Open Road. By using a road energy balance model, a 24 hour prediction of road surface temperature and condition could be transmitted to HAs in the early afternoon based on noon measurements from the road sensors and weather forecast models. A forecast graph, from noon to noon the next day, for a road sensor in each climate zone also enabled the timing of when the road surface temperature was likely to fall below freezing to be seen. As the road sensor data was recorded each hour this was then plotted on the forecast graph to give the engineer feedback as to the accuracy of the forecast. If the road surface temperature forecast was more than 2°C different to the actual temperature and the road surface temperature was below 5°C then the forecast graph would be updated. Secondly, forecast thermal maps were calculated depending on the air stability (extreme, intermediate or damped) for each climate zone. The minimum road surface temperature was forecast and maps of the road network color coded to show which roads were likely to fall below zero. Thirdly, the National Ice Prediction Network (NIPN) was developed to enable weather forecast providers to communicate with any sensor manufacturers instation software and provide additional features such as weather radar and satellite pictures.

It soon became apparent that most HAs would prefer their RWIS to be maintained and monitored by the system suppliers. Hence, the concept of the "Bureau" was developed and the NIPN was then complete.

This system has spread across the whole of the UK and around the rest of the world. When it was developed it was considerably more advanced than any other weather forecast user product but the concept was limited by the speed of computers and communications. In the mid-1980s it took 30 minutes to run the Fortran energy balance model for one location. The modems used to collect the road sensor data often took several minutes to connect to each outstation and were unreliable and slow. The fixed climate zones and thermal map types were necessary simplifications to make the system usable in real-time.

During the 1990s and up to the present day the NIPN – RWIS has speeded up communications and the Internet has been used to access the sensor and forecast data. However, the basic system has not changed and is dated and in need of regeneration.

16.3. Next generation road weather information systems: XRWIS

The RWIS developed in the UK in the 1980s has provided highway engineers with reliable road weather forecasts for almost 20 years. The system architecture provides the highway engineer with site specific forecasts for normally one sensor site in each climate domain. The highway engineer then has to convert this information into a decision as to which salting routes need to be treated, at what time and with how much salt. This is a difficult decision as the engineer may have anything up to 48 salting routes as in Devon. In Birmingham forecast data for one outstation is used to decide which of the 26 routes need salting. This can lead to considerable problems at times of severe winter weather [THO 05].

XRWIS is a simple new intuitive concept (Figure 16.1) that provides a separate road weather forecast for each salting route and gives an initial prediction as to which routes require salting, why they require salting, at what time and with how much salt. A visualization is given in Figure 16.2d – simple "traffic light" colors (red=salt, orange=standby and green=no action). As can be seen in Table 16.1 the expected minimum road surface temperature is given for each salting route together with the optimum time for the salting of the red routes. This information can be presented to the highway engineer for all salting routes in a region or split into divisions as required. The engineer can then "click on" a colored route to see a forecast graph for the coldest point on that route and forecast the road temperature and condition as shown in Figure 16.2. The front-end map will be updated as required by the user.

On many nights all the routes will be red and XRWIS will give the time, reason and salting requirements for each route. On other nights all the routes will be green and no action will be necessary. On several nights a combinations of colors will appear which will reflect a marginal night, thus signaling that the system will need to be monitored through the night. The thresholds for color coding can be agreed with each HA when the system is installed and fine tuned over time by using a self-learning neural network approach.

Ideally the route-based forecasts will be driven by a mesoscale weather forecast model which typically operates with a 10 km grid. The relevant grid point will be chosen for each salting route – or a combination of grid points. This will ensure that the best forecast is made available for each route.

The secret to the success of XRWIS for route-based forecasting is the construction of a geographical database for each salting route [CHA 01a, CHA 01b]. A sky-view factor survey (or thermal geomatic survey if it is combined with thermal mapping) is carried out to provide the proportion of visible sky at 20 m intervals along the route. The sky-view factor controls the energy balance of the road surface

(1=open sky; 0=tunnel) and measures the presence of buildings and trees and other topographic features [CHA 02, CHA 04]. The database also contains other geographical and road data (Figure 16.1) for each 20 m stretch (latitude, longitude, altitude, slope, aspect, road construction, thermal map residual temperature, land-use and traffic volume)

XRWIS then combines the mesoscale weather forecast with the microclimate of each salting route to predict whether or not the route needs treating. Each night is taken on its own merits and there is no need for fixed climate zones and restrictive thermal map types. An energy balance model (IceMiser) is run for every 20 m of road and predicts the road surface temperature and condition every 20 minutes. An additional feature shows a "league table" of salting routes on a given night – ranking the routes from the one that needs salting first to the one that needs salting last. This thermal ranking of routes is very important and will reflect the geography of each route as well as the predicted weather conditions on each night.

In snow situations XRWIS will show the expected snow accumulation for each salting route and the required timing and amount of salt required. Alerts for the use of snow ploughs and snow blowers will be given as an additional layer of information.

The new thermal geomatic surveys (TGS) also enable more efficient salting route optimization and could lead to future developments in selective salting of routes and dynamic routing.

16.4. Verification

During the winter of 2004/05 a trial of XRWIS route based forecasting was initiated for Devon County Council. The trial involved forecasting whether or not six salting routes required treating by using the simple "traffic light" color coding, on a night by night basis. Eight road sensor sites were chosen for verification. A Geographical Information System (GIS) database for each salting route was constructed from a thermal geomatic survey of each route. The UK Met Office supplied the weather forecast data which was based on the 0600 and 1200 run of their mesoscale model. Each salting route had a separate forecast. XRWIS predictions were compared with Devon's sensor data from 1st December 2004 until 30th April 2005 – a total of 152 nights (the data for some nights was missing but at least 144 nights were available for analysis). The trial involved the calculation of road surface temperature for more than 12,000 locations. The forecast minimum road surface temperature for the nearest point to the 8 sensor sites was extracted for each night and compared to the minimum observed road surface temperature, and the results for the 1,200 run are presented in Table 16.2. Also, the average minimum road surface temperature (typical) for each salting route was calculated and compared to the minimum road surface temperature for each sensor site for 121 nights (1st December 2004 – 31st March 2005) and the results are given in Table 16.3. Devon salted their eight sensor sites a total of 329 nights but only on 220 nights did the road surface temperature fall to zero or below. XRWIS predicted that the sites should have salted on 251 nights – which would have meant 78 fewer salting runs.

This trial has been a great success in showing the benefits of the new XRWIS paradigm for predicting which salting routes should be treated. The "typical" (realistic) forecast, which is based on the average forecast road surface temperature for each route, is probably the best approach in that it considers the entire route and is based on more than 2,000 model runs per salting route. Forecasting for the sensor site alone can be misleading depending upon the location of the sensor site.

The XRWIS forecasts did not use road sensor data to initialize their forecasts. This means that reliable forecasts for each salting route can be made without the use of sensor data. The sensors are only used for verification. This new comprehensive route-based forecast approach to road weather forecasting will revolutionize decision making in winter maintenance over the next few seasons.

16.5. Conclusion

XRWIS presents a new intuitive route-based prediction system that gives the highway engineer all the information required to make the correct salting decisions. The delivery is via the Internet with password protection and the geographical database can be built very quickly and economically to enable the microclimate of each salting route to be compiled. The use of virtual outstations and the latest mesoscale weather forecasting models will ensure more accurate salting decisions – reducing both costs and delays as well as improving safety.

The most important potential benefit of XRWIS is that it will lead to more uniform decision making across a region or country. Currently the highway engineers make their own decisions as to which routes to salt based on their own RWIS installation and experience. XRWIS will potentially use standards which are agreed across a region so that every highway engineer in a region would potentially make the same decision for the same weather conditions. The experience of the highway engineers can be built into the algorithms that color code the salting routes and estimate the time of salting and the amount of salt to be used.

It will soon be possible to simultaneously forecast for all the 3,500 salting routes in the UK in less than an hour. This information would then be disseminated to the relevant regional highway engineers who would get a picture of the predicted nights weather across the whole country as well as in their own region. In the event of severe winter weather it would be possible to monitor the forecasts nationwide and share updated information across the country. For example, during the severe winter weather of 28th January 2004 [THO 05], when a squall line developed in front of a cold front bringing Arctic air down from the North, the information on the severity of the event could have been forwarded to the highway authorities in the South several hours in advance.

Like any new paradigm, XRWIS will continue to use the best parts of the old RWIS paradigm. Indeed, all the components of the old paradigm will still be there. XRWIS builds on the current paradigm to give the highway engineer many more layers of information to help them make the correct salting decisions. The intuitive front-end visualization of which routes need to be salted on a given night – synthesizes the geography and microclimate of each salting route with the latest mesoscale weather forecasts. XRWIS route-based forecasting provides all the information that the highway engineer needs.

With increased concern about litigation, highway engineers need to be proactive to ensure that they have the best technology available to them. HA "are under a duty to ensure, so far as is reasonably practicable, that safe passage along a highway is not endangered by snow or ice" (Railways and Transport Safety Act 2003). XRWIS will ensure that the latest cutting edge technology is being applied to the problem.

16.6. Bibliography

[CHA 01a] CHAPMAN L., THORNES J.E., BRADLEY A.V., 2001a. Modelling of road surface temperature from a geographical parameter database. Part 1: Statistical, Meteorological Applications, 8: 409-419.

[CHA 01b] CHAPMAN L., THORNES J.E., BRADLEY A.V., 2001b. Modelling of road surface temperature from a geographical parameter database. Part 2: Numerical, Meteorological Applications, 8: 421-436.

[CHA 02] CHAPMAN L., THORNES J.E., BRADLEY A., 2002. Sky-View Factor Approximation using GPS Receivers, Int. J. Climatol, 22, 615-621.

[CHA 04] CHAPMAN L., THORNES J.E., 2004. Use of neural networks to improve GPS Sky-View Factor Approximation, Journal of Atmospheric and Oceanic Technology, 21, 730-741.

[THO 77] THORNES J.E., WOOD L., BLACKMORE R., 1977. To salt or not to salt? New Scientist, 73, 326-328.

[THO 83] THORNES J.E., SUGRUE G., OSBORNE R., 1983. Thermal mapping of road surface temperatures, Physics in Technology, 14, 212-213.

[THO 91] THORNES J.E., 1991. Thermal Mapping and Road Weather Information Systems for Highway Engineers, in A.H. Perry & L.J. Symons (eds.), Highway Meteorology, E&FN Spon, London, 39-67.

[THO 00] THORNES J.E., 2000. Road Salting: An International Cost/Benefit Review, Proceedings of the 8th World Salt Symposium, Elsevier, Amsterdam, 787-792.

[THO 05] THORNES J.E., 2005. Snow and Road Chaos in Birmingham January 28th 2004, Weather, 60, 146-149.

Salting route ID	Typical minimum temperature (°C):	Treatment to be completed by:
6	–0.5	1700h
8	–1.4	2000h
9	–0.7	2000h
3	–0.1	2300h
4	–0.1	2300h
7	–0.1	0400h
1	0	dry
11	0	dry
2	0.2	n/a
5	0.3	n/a
10	1.2	n/a

Table 16.1. *Example XRWIS action summary (example used is visualized in Figure 16.2d)*

Site forecasts	Rundlestone	RD&E	Exton Camp	Little Stone	Haldon Hill	Cadbury Cross	Craze Lowman	Ashmill	Mean
Number of nights	151	152	147	152	147	148	152	144	149
Percent correct	89.4%	91.4%	94.6%	88.8%	91.2%	91.2%	88.8%	82.6%	90%
Bias	0.85	1.73	1.16	1.85	1.12	0.97	0.82	1.6	1.3
Miss rate	0.23	0.07	0.09	0	0.2	0.2	0.32	0.06	0.15
False Alarm Rate	0.04	0.09	0.05	0.13	0.07	0.05	0.04	0.21	0.09
Pierce Skill Score	0.73	0.84	0.86	0.87	0.73	0.75	0.64	0.73	0.76

Table 16.2. *A summary of the forecast quality for each site using the 1,200 forecast run*

Typical forecasts	**Rundlestone**	**RD&E**	**Exton Camp**	**Little Stone**	**Haldon Hill**	**Cadbury Cross**	**Craze Lowman**	**Ashmill**	**Mean**
Number of nights	121	121	121	121	121	121	121	121	121
Percent correct	80.0%	88.4%	88.3%	90.0%	90.1%	89.3%	86.0%	85.1%	87%
Bias	0.51	1.56	1.26	1.65	1.08	0.85	0.97	0.76	1.1
Miss rate	0.49	0.125	0.17	0	0.17	0.14	0.24	0.38	0.21
False Alarm Rate	0.01	0.11	0.1	0.13	0.07	0.05	0.1	0.06	0.08
Pierce Skill Score	0.50	0.77	0.73	0.87	0.76	0.81	0.66	0.56	0.71

Table 16.3. *A summary of forecast quality Typical forecasts using the 0600 forecast run*

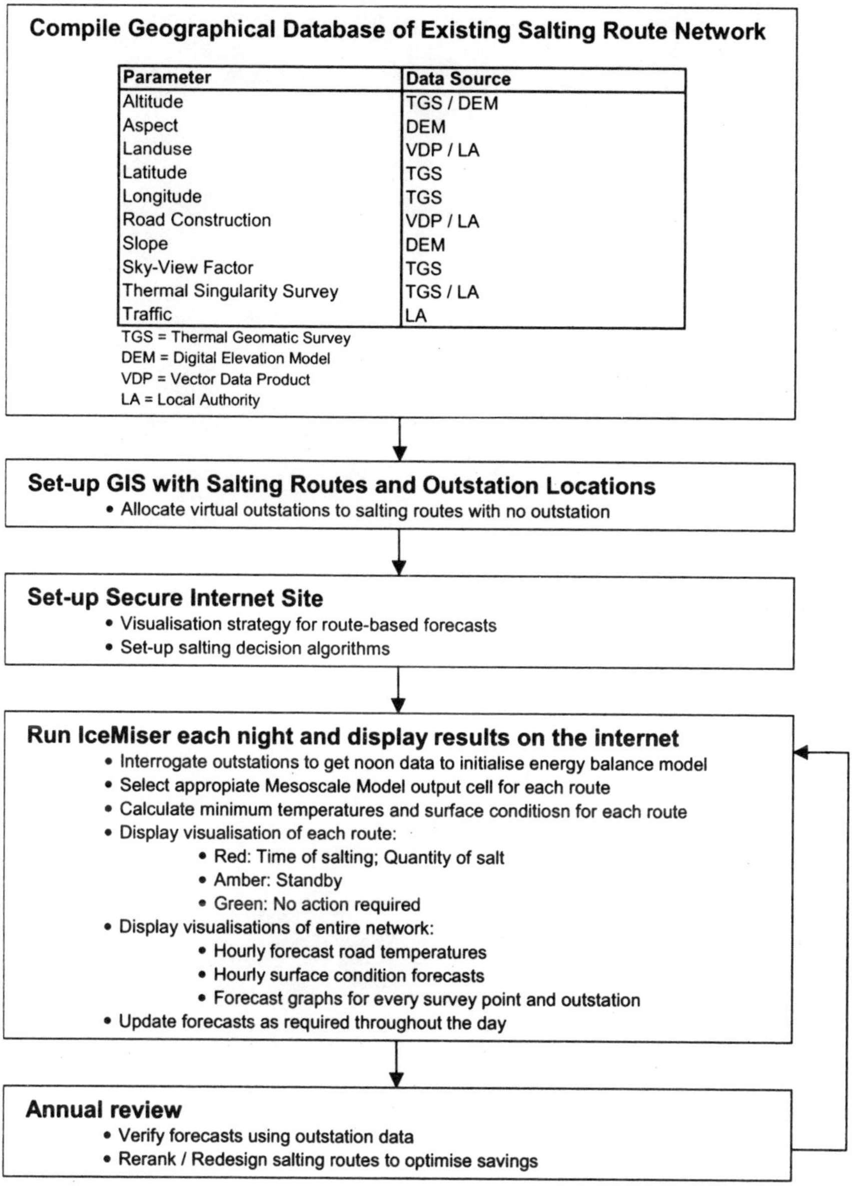

Figure 16.1. *Flow diagram showing the procedure in developing XRWIS*

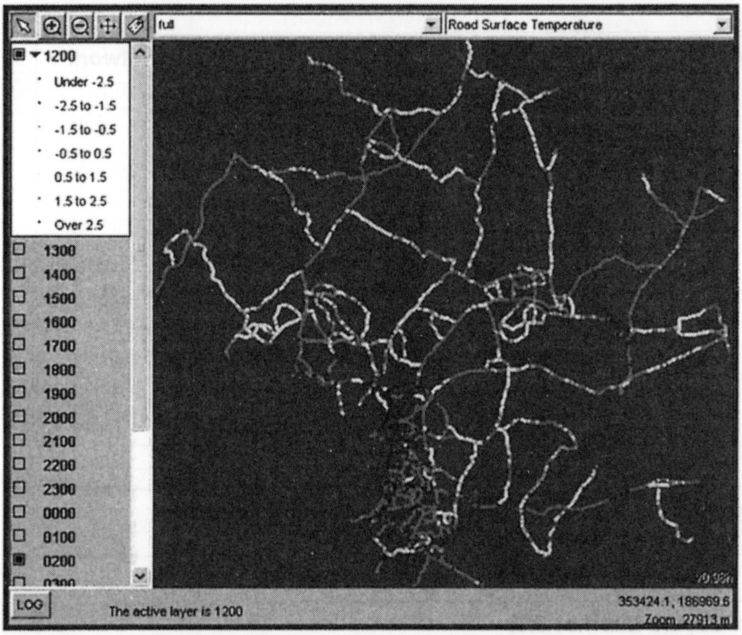
full
Road Surface Temperature
1200
Under -2.5
-2.5 to -1.5
-1.5 to -0.5
-0.5 to 0.5
0.5 to 1.5
1.5 to 2.5
Over 2.5
1300
1400
1500
1600
1700
1800
1900
2000
2100
2200
2300
0000
0100
0200
LOG
The active layer is 1200
353424.1, 186969.6
Zoom 27913 m

a)

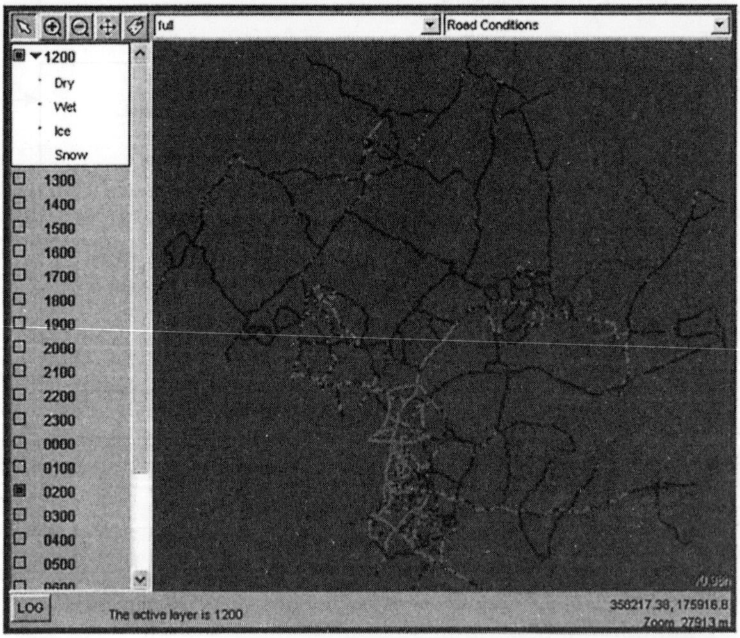
full
Road Conditions
1200
Dry
Wet
Ice
Snow
1300
1400
1500
1600
1700
1800
1900
2000
2100
2200
2300
0000
0100
0200
0300
0400
0500
LOG
The active layer is 1200
358217.38, 175916.8
Zoom 27913 m

b)

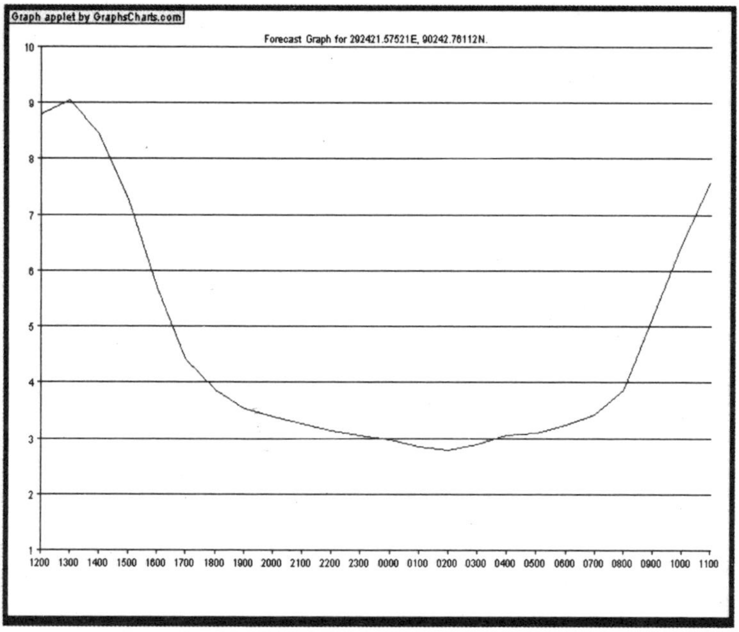

c)

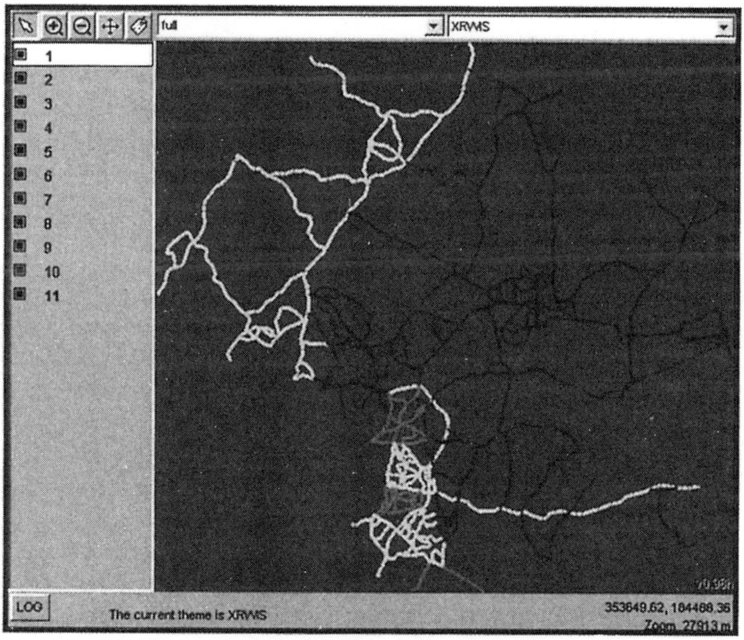

d)

Figure 16.2. *Conceptual visualization of the new XRWIS paradigm showing: a) road surface temperatures, b) road surface conditions, c) forecast graph and d) "traffic light" salting routes*

Part 4

Climate-related Applications

Chapter 17

The Use of GIS in Climatology: Challenges in Fine Scale Applications: Examples in Agrometeorological and Urban Climate Studies

17.1. Aim and context

The use of GIS in climatology is at present led by *two different fields of applications*:

– produce and display *global to regional scale* spatially interpolated climatological fields for climate change and impact studies. This field mainly uses controlled climatological information from national meteorological offices, which is normalized according to international standards. Products may be used down to local scales according to top-down disaggregation procedures. Topographic parameters from DEMs and/or satellite remote sensing data are the most common adjuncts in the spatial interpolation methodologies. This is the main stream of studies within the COST 719 project;

– handle (i.e. archive, homogenize, interpolate, display, etc.) the *fast growing amount* of locally acquired and diversified climatological and air pollution data, which is acquired at *fine scales* for the knowledge and prevention of climate or air pollution related *risks*. There are numerous initiatives without coordination, involving various scientific communities and local actors as well.

Chapter written by Claude KERGOMARD.

17.2. GIS challenges in fine scale applications

We mainly consider here four of them.

17.2.1. *First challenge: handle, homogenize and archive "atmospheric information"*

This includes data from various types and origins and data from *punctual measurements* including:

– standard meteorological data from national networks;

– specialized networks for specific applications: agrometeorology, urban rain gauges, air pollution networks, etc.;

– unconventional field measurements from temporary networks and mobile systems;

and imagery from *remote sensing* systems including:

– satellite or airborne remote sensing (if available at useful resolution);

– ground-based remote sensing systems such as pluviometric radar, wind profiling Doppler systems, lidar, sonar, etc.;

The common *geo-referencing* of such various data is often a time-consuming task!

17.2.2. *Second challenge: handle, synthesize and prepare "geographical information"*

This information is requested for spatializing and/or physical modeling of fine scale climate/air quality fields:

– topographic variables from DEM: altitude, slope, aspect, etc. Topographic variables provide *continuous quantitative fields*;

– surface parameters interacting with the atmosphere: vegetation, soil properties, land cover and land use, etc. At a finer scale, surface properties become more significant and the geographical information is more *qualitative and discontinuous*.

17.2.3. *Third challenge: spatial interpolation of climatological/air quality data*

This is strongly challenged by the irregular distribution and/or insufficient density of field measurements. Spatial interpolation in fine scale applications frequently applies to individual or extreme situations rather than time-averaged climatological fields.

Therefore, geostatistical procedures (kriging) often fail in providing accurate climatological or air quality fields. GIS methodologies using independent predictors have to deal with the aforementioned weight of air-surface interactions and with the complexity of surface descriptors.

Spatial interpolation procedures often have to *be adapted to each case*, lacking reproducibility.

17.2.4. *Fourth challenge: GIS-based spatial interpolation*

These procedures always have to be confronted to the reality of *physical processes* acting at different time and space scales:

– large scale external forcings;

– "fluid" processes including air mass advection, vertical mixing (convection), local breezes;

– air-surface interactions including surface radiative and heat budget, mass and dynamical energy exchanges.

The weight of these different processes may vary significantly for different meteorological situations. The weight of air-surface interactions at local scale through the radiative and heat budget is emphasized during "radiative" situations (clear sky, weak wind or lack of it). A special interest is devoted to this kind of situation (frost hazard, urban heat island). Is there competition or complementarity between *spatial interpolation* and *mesoscale models*?

17.3. Examples of application in agrometeorology

This chapter deals with two different fine scale applications of GIS, which are illustrated by the works of four young researchers' PhD Thesis in the last few years:

– *Agroclimatological hazards*, especially spring frost hazard in the most delicate vineyard and fruit orchard with high economic values (see [SAI 99, MAD 04, QUE 02];

– *Urban climate*, especially urban heat island spatialization and processes (see [CHA 03, LON 03]).

17.3.1. *Spring frost hazard in the Champagne vineyard*

A traditional approach of frost hazard is provided by *probability calendars*, first defined at Grenoble University and applied to frost hazard in the Champagne vineyards by Marie Françoise de Saintignon in the 1990s [SAI 90]. To spatialize something like a frost hazard in a vineyard, we have to deal only with a few days in each year, a few days where the probability of frost remains small and the impact of these days is clearly related to the phenology of wine, which is different for chardonnay, pinot meunier, pinot noir, etc. This calendar of probability shows us that we have to discuss which kind of temperature we are interested in. This is not the standard temperature using a 2 m shelter, but it is the actinometry temperature at 50 cm without shelter and a strong soil indirection.

A first application by Madelin [MAD 04] deals with frost risk in the Champagne area in 2003: there were a lot of damages with a big percentage of grapes destroyed by frost within a few days in April. What can we use for such a question? We can use a special kind of network, which gives measurements of standard temperature, minimal temperature in the shelter and actinometric temperature. All the meteorological stations were located either on the top of the slope, at the bottom of a slope or on the flat surfaces. We have to consider that very carefully: this is the initial material. We also had to use a land cover database and an altitude database to build a digital elevation model.

This is a traditional GIS procedure for interpolating data as did Madelin, by using different layers and co-registration and co-use of these different predictant and predictors. In her first attempt, the land cover was used in a very simple manner: Malika only used a mask for vineyard areas, arguing that it was not necessary to interpolate data in other kinds of land cover. Therefore, she produced maps of minimal temperature for selected days. The main problem is that, for very similar meteorological situations with only slight differences in wind direction and wind force, she had very different estimations. In this situation, the local factors including the relative altitude of the station on the slope, using the north and south exposure were the main estimators and the regional tendency had only a limited importance.

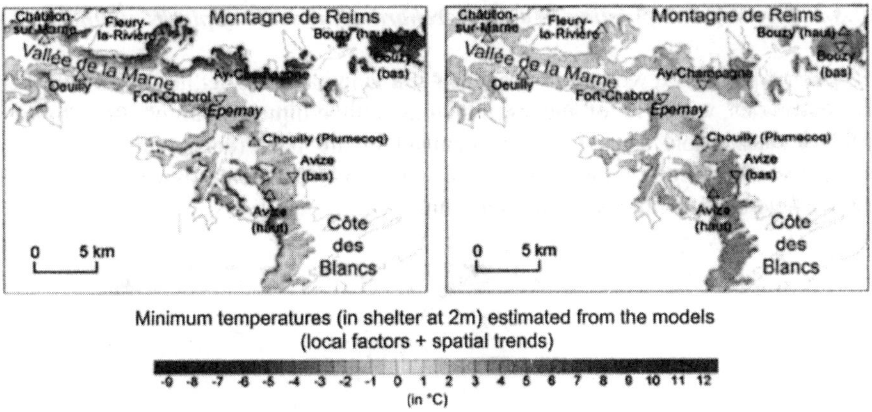

Figure 17.1. *Spatialization of frost hazard in the Champagne vineyard (from [MAD 04]): minimal (night) temperatures estimated from a regression model by using observations, topography from a DEM and land cover*

– The left image (17/03/2003) shows a rather good regression model (64.2% of variance explained) during a clear sky and weak wind weather situation. The local features due to topography are emphasized.

– The right image (20/03/2003) shows a worse regression model (only 39.3% of variance explained) and a regional distribution of temperatures due to the effect of a synoptic wind.

Her second attempt was quite different. The regional tendency is more significant, so for quite similar situations we have two very different regression models. This leads of course to very different maps of minimal temperature in this area. In this case, Madelin also used this terrain data in competition or in cooperation with the physical modeling experiment in the area: their results validate the physical model but also the physical model brings a lot of information about more precise processes (including, for example, flow along the terrain), especially vertical flow along the terrain when the low temperature can form along slopes with different exposure.

This is what we can establish as something like a state of the art in agro-climatological applications of GIS for such areas. However, in the case of Champagne, the problem of land cover was completely avoided because Madelin only worked on the vineyard land use at medium scale.

What happens when we go to finer scales?

17.3.2. *Towards interpolation in a fruit orchard at the scale of pieces of land*

Quénol [QUE 02] worked in Provence at a land parcel scale: he had to consider discontinuities, with a lot of hedges, including a three-dimensional representation of cold air distribution, here with an abatement for the TGV (very high speed train) across the area. With all these continuities and discontinuities: it is very difficult to build a field of temperature in such a terrain.

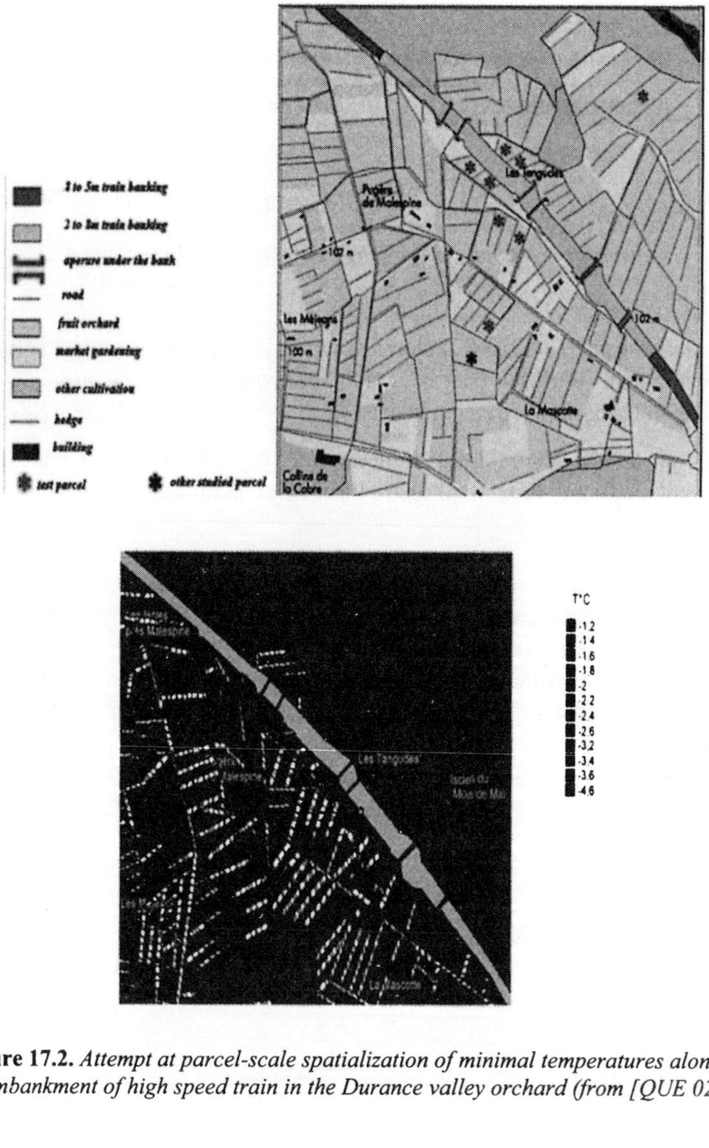

Figure 17.2. *Attempt at parcel-scale spatialization of minimal temperatures along the embankment of high speed train in the Durance valley orchard (from [QUE 02])*

The spatialization model requires the use of both vector and raster GIS approaches to take into account the discontinuities in land use and windbreak hedges effect on the continuous temperature field.

Quénol did good work and proposed a method which uses a combination of raster and vector systems, thus producing a temperature field which is typical of what can be produced in such terrain. He showed a way for further studies and further meanings for when we have to discuss the use of GIS in climatology at very fine scales.

17.4. Urban studies examples

Let us now analyze urban studies and urban climate studies examples.

17.4.1. *Urban heat island in the Lille metropolitan area [CHA 03]*

In urban climate studies, the heat island is traditionally obtained interpolating atmosphere and terrain parameters but, for an urban area like Lille (Northern France), we have measurements only for one or two days at a fixed hour, around 6h in the morning. All these measurements only give us a preview of the urban heat island detail.

How is it possible to spatialize such punctual data?

We have of course to deal with what the city really is and what is the relationship between the temperature measured and the city. In a very simple application [CHA 03], we only used a two-dimensional view of the city and we had to ask ourselves what the measurement is and in which manner this measurement of temperature here is only affected by grass and trees in a park, water or city, what the size of the environment of the stations we have to take into account for spatial interpolation is. Therefore, when we can produce a map by using an urban database from Lille Metropole Urban Community, we cannot really know what it tells because we had to choose a space and we cannot deal in a simple manner with a multi-scale system, so that this study will be compared with other works which were made in Marseilles.

17.4.2. *GIS-based analysis of urban fabric for use in urban climatology [LOW 03]*

This work took place within a sub-project of a big experiment called "ESCOMPTE Urban Boundary Layer". During this experiment, we had to manage urban databases in order to provide significant values about the city. What is significant when we try to work with a city and climate interaction?

The most interesting problem is that the city is not two-dimensional, it is three-dimensional and we have to deal with the geometric spacing of the buildings, because this geometry is clearly related to the heat budget of the city, including the well known street canyon effect, which traps energy, but it is also related to dynamical interactions.

How is it possible to deal with very different urban fabric types in a city like Marseilles? Marseilles is a rather complex city with pronounced topographic features, and a mix of dense urban fabric in the center, with typical small individual areas of buildings and high-rise residential complexes. So we used the "BD Topo" database (from the Institut Géographique National), which includes surface types, a digital elevation model, a fine description of buildings' height. How can these layers be used for climatological studies?

It was Nathalie Long [LON 03] who first tried to use together elevation and building heights to produce a map with a grid mesh which is compatible with the modeling purposes in order to provide significant data. For example, we produced a grid mesh of 200 x 200 m comprising surface densities of water, and of vegetation, built surface, building morphology, heights, volumes, perimeters, compactness, building distribution, average spacing, numbers of building in a mesh, street average widths, aspect ratio for the streets and, from these descriptions, we used dynamical parameters together like rugosity lines and several other descriptors of rugosity.

We obtained for a map of Marseilles for each kind of urban fabric including the center, the periphery districts of the mid-19th century and the individual housing center. From the qualitative discontinuous description of the city we got quantitative and grid information for modeling.

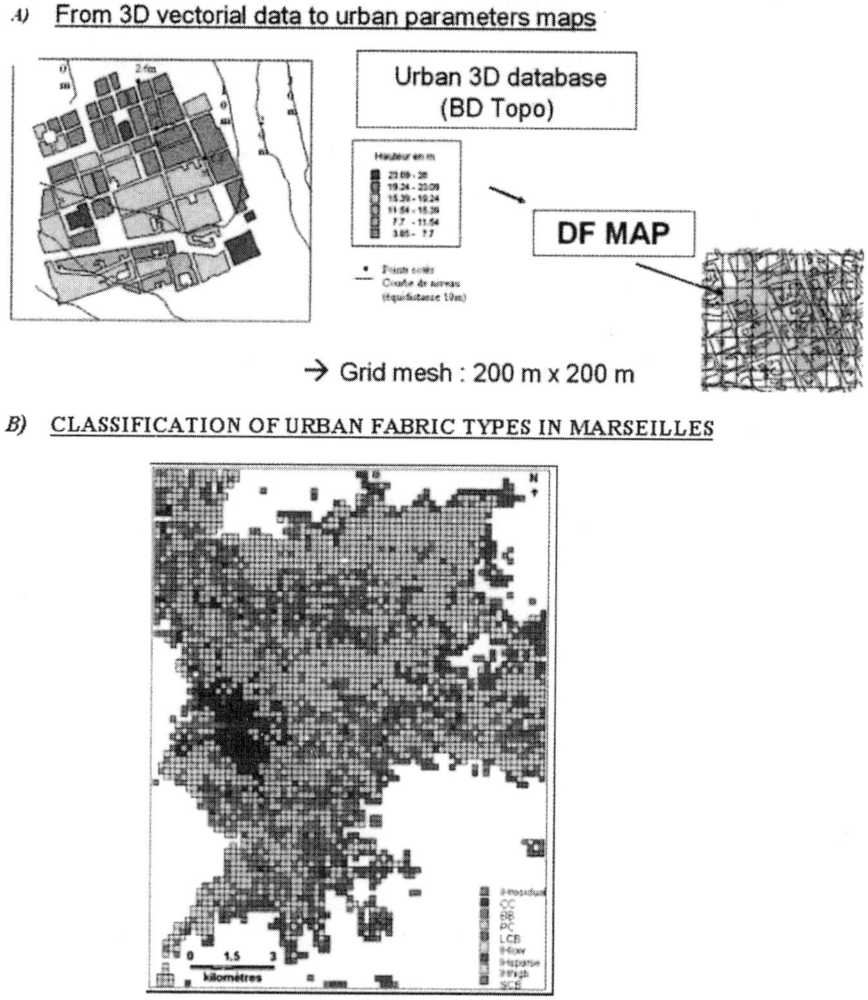

Figure 17.3. *GIS-based analysis of the urban fabric for use in urban climate studies (from [LON 03])*

This process was compared to other ways to provide information for modeling, which is a comparison of all our data with the more traditional land cover information used by Meteo France models (Corine Land Cover which, in Marseilles, only considers two main classes: continuous and discontinuous urban areas). We obtained much more detailed and quantified information: as a complement, a classical supervised classification of a standard Spot image provided a 20-m resolution distribution of the surface, but without more quantitative information. Our layers were used together with the data of a meteorological network in the city. This meteorological network is mainly related to the distance to the sea and we can use a lot of information from urban descriptive parameters.

In this project, GIS was also used to manage very different types of information. For example, we had to manage small images from an infrared airborne camera which provided the temperature distribution of various places in the city and we had to find relationships between this distribution of brightness temperatures and the city parameters. It enabled us to see, for example, that in a council estate in the northern parts of Marseilles a big building showed a temperature distribution with a wide range, a wide standard deviation, but not very high temperatures (except for the very small area including the main building faces) and that in other areas we found higher temperatures, but also more homogenous ones.

From this example it is possible to understand the big problem of fine scale climatology: *the surface is more and more diverse, more and more complex as we move towards the finer scales*. This can no longer be simply interpreted as continuous fields, as it was done in traditional climatological studies, and geostatistical procedures are not convenient. So we have to find more and more sophisticated uses of GIS but it is probably not yet possible to create very reproducible traditional models like those used in COST projects for other states at less finer scales.

17.5. Acknowledgements

Thanks are due to the COST 719 team for the final conference in Grenoble. The kind help of Pierre Dumolard in transcribing my talk is gratefully appreciated.

17.6. Bibliography

[CHA 03] CHARABI Y., "L'ilôt de chaleur urbain de la métropole lilloise: mesures et spatialisation", PhD Thesis, Lille 1 University, 2003.

[LON 03] LONG N., "Analyses morphologiques et aérodynamiques du tissue urbain. Application à la micro-climatologie de Marseille durant la campagne Escompte", PhD Thesis, Lille 1 University, 2003.

[MAD 04] MADELIN M., "L'aléa gélif printanier dans le vignoble marnais en Champagne. Modélisation spatiale aux échelles fines des températures minimales et des écoulements d'air", PhD Thesis, Paris 7 University, 2004.

[QUE 02] QUÉNOL H., "Climatologie appliquée aux échelles spatiales fines. Influence des haies brise-vent et d'un remblai ferroviaire sur le gel printanier et l'écoulement du mistral", PhD Thesis, Lille 1 University, 2002.

[SAI 99] (de) SAINTIGNON M.F., "Probabilité de gel thermique dans le vignoble champenois, unpublished report; Laboratoire de la Montagne Alpine, Joseph Fourier University, Grenoble, 1999.

Chapter 18

Climate Impact on the Winter Land Use and Land Cover Management in Brittany

18.1. Introduction

In intensive agricultural regions, the land use and land cover have an important impact on the water resources quality. An accurate assessment of the spatial and temporal variation of winter vegetation represents a key indicator of water transfer processes, which is essential for controlling land management and helping local decision making.

The area of study is a watershed located in the south of Brittany in France. With a surface of 500 km², this watershed provides potable water for 80,000 people, thus making the water quality on this area a prior sanitary task for the local managers. That is why they decided to build a Geographical Information System (GIS) in order to be able to manage the dynamics of the land cover changes and take optimal decisions in the problematic areas. The follow-up of the land cover is realized with optical remote sensing data (SPOT, IRS-LISS) and integrated in the GIS at a field scale.

The local managers focus especially on the land use and land cover in winter that represents an intermediary period for the main crops in this region which are characterized by a relative warm and rainy season. The presence of bare soils during this period, which accelerate the flux transfers on the water resources, can thus have a significant influence on the concentration of polluting matters. The location of

---

Chapter written by S. CORGNE, H. QUÉNOL, O. PLANCHON and T. CORPETTI.

these "hot spots" appears fundamental for the local decision makers. Therefore, a predictive model was built:

– to identify the factors that motivate the land use and land cover changes between winters; and

– to spatially predict the location of the land cover for the next winter in order to anticipate specific environmental actions.

Four driving factors have been identified and statistically validated for modeling the land cover in winter: "bare soils occurrence", "field size", "main crop precedent" and "land cover evolution". The predictive model based on the Dempster-Shafer Theory (DST) and the Dezert-Smarandache Theory (DSmT) made it possible to perform correct predictions for the next winter at a field scale of respectively 74% for the land cover with DST and 81% with the DSmT.

The objective of this chapter is to take in account the impact of the climate on this prediction. The watershed is characterized by a relative homogenous pluviometry, however, spatial variations can locally influence the land cover implantation. We quantify and integrate here the factor "climate" in the model relative to the others factors and measure its pertinence for the land cover prediction in winter at a field scale.

18.2. Climate characteristics of the study area

18.2.1. *Site description*

Brittany and all the West and North-west of France are usually subjected to the Atlantic weather disturbances throughout the year, but their frequency and intensity are subjected to a seasonal variability, which directly determines the precipitation regimes. At small scales, topographical effects (the highest hills of Brittany reach between 300 and 400 m) influence the effect of the circulation type on precipitation. During early winter (December and January), the Westerlies are strong and the pressure systems have a dominantly east-west orientation (high zonal index). Heavy precipitation is recorded on the European Atlantic fringe: December and January are the two rainiest months at Lorient in Brittany. Starting from February, the lower zonal index is connected to the weakening of the westerly circulation and the higher occurrence of the meridional circulations [BAR 87]. The highest occurrence of the meridional circulations is recorded in spring (April and May, [LEJ 83]). The precipitation amounts usually decrease in spring on the western fringe of Europe, but the development of low geopotential heights has strong effects on cyclogenesis and precipitation, depending on the longitudinal position of the troughs [MOR 90, PLA 05]. The summer is subjected to a higher occurrence of westerly weather patterns, but the weather disturbances tracks are high in latitude. Relatively low precipitation amounts are recorded on the European Atlantic fringe (at Lorient, June-July-August: 15.6% of the annual amount), but heavy precipitation can occur depending on the weather disturbances conditions. The westerly circulation strengthens in autumn, involving an increase of precipitation in Western Europe.

18.2.2. *Meteorological information*

18.2.2.1. *Meteorological data*

The climatic data used in this study is the daily precipitation recorded during the period of 2000-2004 at 3 meteorological stations (Lorient, Plouay and Plouray), which are distributed in the Scorff watershed (Figure 18.1). The station of Lorient is the nearest to the Atlantic Ocean. The stations of Plouay and Plouray are located in the central (Plouay) and in the northern part (Plouray) of the Scorff watershed, close to the Black Mountains (Table 18.1). These stations belong to the French national meteorological office (*Meteo France*). In addition to this precipitation data between 2000 and 2004, the monthly precipitation data for the period of 1971-2000 has been used as a reference to define the precipitation anomalies.

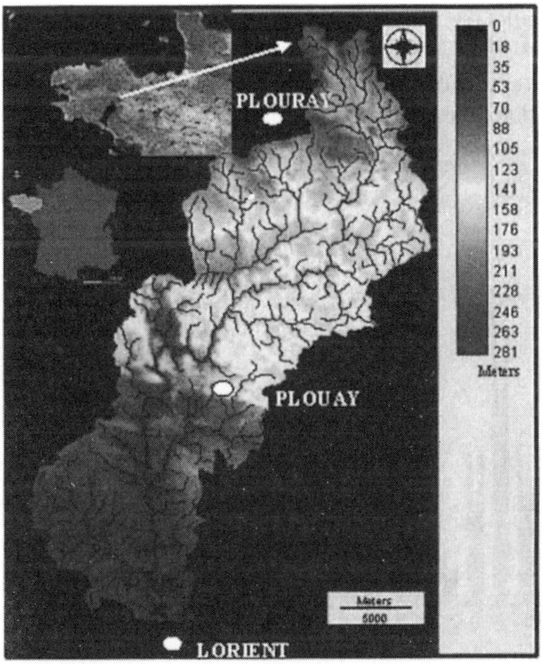

Figure 18.1. *Localization of the area of study*

Station	Latitude	Longitude	Altitude	Distance/coast
Lorient	47°46'N	03°27'W	42 meters	5 km
Plouay	47°56'N	03°19'W	138 meters	22 km
Plouray	48°07'N	03°26'W	205 meters	44 km

Table 18.1. *Geographical coordinate*

18.2.2.2. *Space-time variability analysis of precipitation on the Scorff watershed using the three reference stations*

Figure 2 shows the comparison between the monthly precipitation amounts for the period of 2000-2003 at three reference stations located in the Scorff watershed (Lorient, Plouay and Plouray). The coastal station (Lorient) was clearly less rainy than the two other (inland) stations. Topographical effects and especially the roughness of the hilly area cause higher precipitation amounts at the inland stations. The most inland station (Plouray) is the rainiest station: Plouray is located close to the southern slope of the Black Mountains, which reach about 300 m on the north-western edge of the Scorff watershed. For the month of December 2000, the precipitation amounts recorded at the stations of Lorient, Plouay and Plouray were respectively 207.4 mm, 338.1 mm and 350 mm. The southern to south-western exposure of the watershed allow to enhance the heavy precipitation due to the W, SW and S circulation types, especially in the hilly inland area.

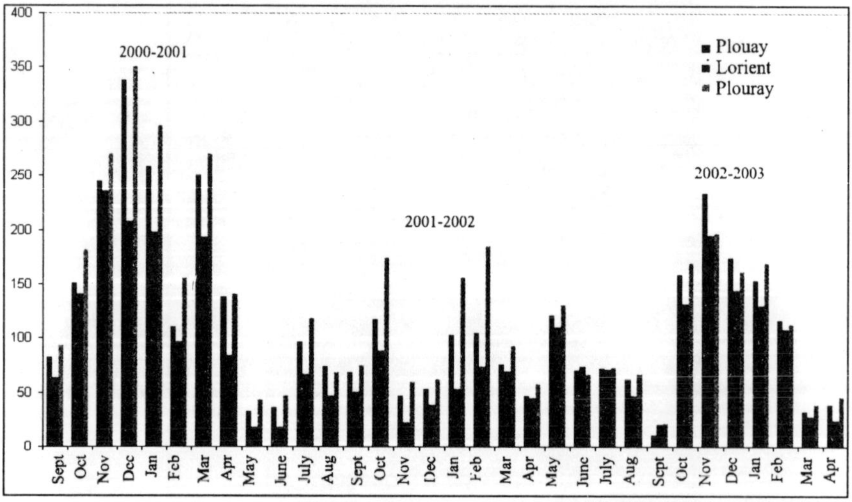

Figure 18.2. *Monthly precipitation at LORIENT, PLOUAY and PLOURAY (histogram)*

18.2.3. Relationships between precipitation space-time variability and land cover management

To analyze the relationships between precipitation space variability and land cover in winter on the Scorff watershed, a Digital Elevation Model (DEM, resolution: 50 meters) was used. The DEM was classified into three classes according to the altitude of the three stations (class 1: less than 100 meters; class 2: 100-170 meters; class 3: more than 170 meters). Three winters were studied here (2000-2001; 2001-2002; 2002-2003). The determination of the land use at a field scale represents a very important preparation to accurately detect the vegetation, it is realised from high and very high resoltion satellite data (SPOT XS; SPOT 5). The resolution of the data makes it possible to categorize "bare soils" and "covered soils" at a field scale [LAM 94, COR 02].

Numerous statistics tools exist to analyze the relations between the explicative variable and the variable to explain (regression, factorial analysis, etc.). A method based on the mutual information between the variables was applied here to evaluate the significance of the DEM on the land cover and land use. This method is based on the Information Theory [SHA 49] and makes it possible to quantify the information relative to an event in relation with the information obtained by the realization of this event. The application of this theory offers several indicators, among them the Kullback criteria [KUL 59] that approximately corresponds to the χ^2 law. It makes it possible to measure the statistical link between the explicative variable (precipitation) and the variable to explain (land cover in winter).The most significant relationship between the explicative variable (DEM) and the variable to explain (land cover: "bare soils", "covered soils") was measured for the winter of 2000-2001 (Table 18.2) which was characterized by a high precipitation spatial variability between the stations and high precipitation amounts for this winter (K = 683.03). On the other hand, when the spatial variability between the stations was relatively weak as for the winter of 2002-2003, the explicative variable (DEM) became less significant (K = 5.54). Application of the Kullback criteria shows that topographical effects (precipitation) have a real statistical relation with the land cover when the precipitation spatial variability between the stations is high. For example, for the winters of 2000-2001 and 2001-2002, we noted a high difference of the land cover between classes 1 and 3 compared with the winter of 2002-2003 where spatial variations of the land cover were weak. Thus, relations between precipitation and the land cover management in winter can be significant when the spatial variability of the precipitation on the Scorff watershed is noted.

Explicative variable (DEM)	**2000-2001 (bare soils)**	**2000-2001 (cover soils)**	**Kullback criteria (K)**
Class 1 (< 100 meters)	2,045 parcels 20.9%	7,751 parcels 79.1%	
Class 2 (100-170 meters)	1,164 parcels 9.63%	10,921 parcels 90.4	**683.03**
Class 3 (> 170 meters)	368 parcels 8.37%	4,030 parcels 91.6%	
	2001-2002 (bare soils)	**2001-2002 (cover soils)**	
Class 1 (< 100 meters)	1,671 parcels 17.1%	8,125 parcels 82.9%	
Class 2 (100-170 meters)	1,458 parcels 12.1%	10,627 parcels 87.9%	**114.61**
Class 3 (>170 meters)	572 parcels 13%	3,826 parcels 87%	
	2002-2003 (bare soils)	**2002-2003 (cover soils)**	
Class 1 (<100 meters)	1,173 parcels 12%	8,623 parcels 88%	
Class 2 (100-170 meters)	1,527 parcels 12.6%	10,558 parcels 87.4%	**5.54**
Class 3 (> 170 meters)	587 parcels 13.3%	3,811 parcels 86.7%	

Table 18.2. *Relationships between space-time variability precipitation and land cover for three winters*

18.3. Impact of the climate characteristics in the land cover prediction model

In intensive agricultural regions, short-term land use and cover changes represent an important key indicator of water transfer processes. The presence of fields with a lack of or little vegetation during the winter increases pollutant fluxes towards the rivers. The accurate assessment of the spatial and temporal variation of winter vegetation covering is essential for controlling land management and helping local decision making. The spatial prediction modeling of winter bare soils is complex due to the intricate interactions between biophysical and socio-economic factors operating at various scales from the field to the watershed. In some cases the uncertainty level can be significant, as far as meteorological constraints or economic and/or political events like subsidy allocation for certain crops are concerned. Therefore, it is necessary to introduce uncertainty in modeling land use and cover changes, especially when high spatial and temporal variability is encountered.

18.3.1. *Presentation of the land cover prediction model*

In order to take in account uncertainty in the data used and in the knowledge of the problem, the Dempster-Shafer theory (DST), which is also named the theory of evidence, was chosen. It introduces uncertainty in modeling, enables the expression of ignorance in the body of knowledge and states that belief in a hypothesis is not necessarily the complement of its negation. In remote sensing, this theory has been applied in multi-sensor data fusion with the main objective of improving the classification by integrating multiple data sources [HEG 00, LED 01]. It has been recently applied for land cover prediction [COR 04] and fusion of vegetation indices [HEG 06]. Detailed applications of the Dempster-Shafer theory can be found in [MER 01]. Despite the well-known advantages of Dempster's rule, DST presents several limitations [DEZ 03] especially when the conflict between the information sources becomes significant. Thus, a new theory named Dezert-Smarandache Theory (DSmT), which can be considered as a generalization of the DST, has been developed in order to manage the conflict between the information sources [DEZ 03].

18.3.1.1. *Presentation of the DST and DSmT*

DST and DSmT are data fusion rules that make it possible to combine different information sources. They can manage uncertain and imprecise data in the modeling process (and paradoxical data for DSmT). The theory of evidence proposed by Dempster was developed by Shafer in 1976 and the basic concepts of this theory have often been exposed in [SRI 90]. A detailed presentation of DSmT can be found in [SME 03], 2003. We present in Figure 18.3 the simplified principles of DST and DSMT. Let us assume that a domain reference, *called the frame of discernment* Θ, is composed of a set of exhaustive and mutually exclusive hypotheses [1]. For our case, we focus on the hypotheses "bare soils" and "covered soils", the uncertainty is represented by the union of the hypotheses (DST) and the paradox with the intersection of the hypotheses (DSmT). The mass function affectation for each information source (pieces of evidence) is generated and supports one of the hypotheses. The sum of all basic probability assignments or masses is always equal to 1 and *m* for the empty set is 0 [2]. All pieces of evidences are then combined with the DST and DSmT data fusion rules [3]. The decision making is calculated with the maximum pignistic probability function [4]. All the results are validated with a classification of a remote sensing data. Thus, for the land cover prediction presented here for the winter of 2003-2004, all the results produced by the model were spatially confronted with the classification produced for the winter of 2003-2004.

<u>**Dempster-Shafer Theory**</u>	<u>**Dezert-Smarandache Theory**</u>
1. Frame of discernment (Θ) Θ = {bare soils (Bs), cover soils (Cs)}	**1. Frame of discernment** (Θ) Θ = {bare soils (Bs), cover soils (Cs)}
– Powerset $(2^\Theta) = \{Bs, Cs, Bs \cup Cs, \varnothing\}$	– HyperPowerset $(2^\Theta) = \{Bs, Cs, Bs \cup Cs, Bs \cap Cs, \varnothing\}$
2. Mass function affectation – $m(\varnothing) = 0$ and $\Sigma\, m(A) = 1$ where $A \subseteq \Theta$	**2. Mass function affectation** – $m(\varnothing) = 0$ and $\Sigma\, m(A) = 1$ where $A \subseteq \Theta$
3. Beliefs, measures and decision making – Credibility (pessimist belief) – Plausibility (optimist belief) – Pignistic (optimized function)	**3. Beliefs, measures and decision making** – Credibility (pessimist belief) – Plausibility (optimist belief) – Pignistic (optimized function)
4. Date rule combination with DST $m(C) \equiv [m_1 \oplus m_2](C) = \frac{\sum_{A\cap B=C} m_1(A)m_2(B)}{1-\sum_{A\cap B=\phi} m_1(A)m_2(B)}$	**4. Date rule combination with DST** $\forall C \in D^\Theta, m(C) = [m_1 \oplus m_2]\ (C) = \sum_{A,B\in D^\Theta, A\cap B=C} m_1(A)m_2(B)$
5. Decision making – Maximum of the pignistic probability	**5. Decision making** – Maximum of the pignistic probability

Figure 18.3. *General principals of the Dempster-Shafer and Dezert-Smarandache Theories*

18.3.1.2. *Change prediction design and results of the land cover prediction with DST and DSmT*

Firstly, the change prediction model consists of determining, analyzing and putting in a hierarchy the factors that motivate the land cover changes between the winters (Figure 18.4). For this step, we used numerous types of information (remote sensing, expert knowledge, statistical analysis, etc.) that made it possible to discriminate four driving changes factors: "bare soils occurrence", "field size"; "main crop precedent" and "land cover evolution". Secondly, mass function affectation was realized for all this "pieces of evidence" or "information sources" that partially support one of the hypotheses of the powerset (DST) or the hyperpowerset (DSmT). For example, for the information source "field size", the statistical analyzes showed that there are more bare soils for the parcels which are more than two hectares. Thus, this piece of evidence will preferably support the hypothesis "bare soils". As the information sources are characterized by a different level of uncertainty or paradox, some mass functions are affected by the union of the hypothesis (DST) or the intersection of the hypotheses (DSmT) when the conflict between the information sources is significant. All information is combined with the data fusion rule and the decision making is realized with the maximum of pignistic probability.

The results with the two theories are exposed in Table 18.3. The DSmT with 81.3% of total good predictions offered better results than the DST (73.8%), but for the two theories the bare soils predictions were quite weak: only 22% for DST and 9% for DSmT. On the other hand, the soils with vegetation were relatively well predicted (81.3% for DST and 91.6% for DSmT). The bare soils prediction appeared difficult to model because of high spatio-temporal changes that depended on multiple factors such as the economical context, climate, agricultural strategies, etc. The climate (precipitation) factor has a statistical relation with the land cover management in winter when there is a high spatial variability of the precipitation. Thus, his integration in the land cover predictive model can contribute to improve the results.

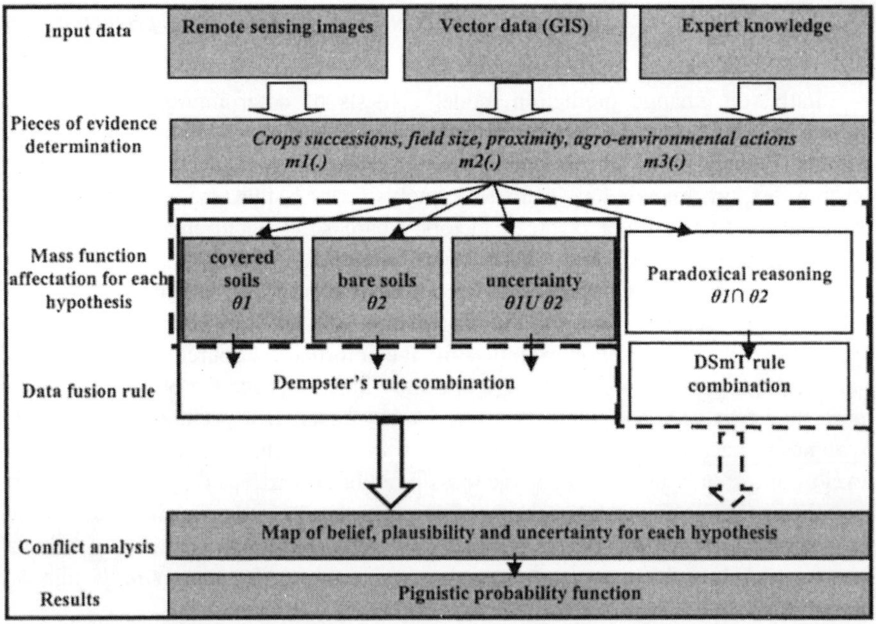

Figure 18.4. *Land cover prediction modeling design*

	Land use in winter 2003/04	**Dempster-Shafer Theory**		**Dezert-Smarandache Theory**	
	remote sensing data (validation)	**Prediction**	**Right prediction**	**Prediction**	**Right prediction**
Bare soils	3,609 ha	5,502 ha	792 ha **(22%)**	2,445 ha	338 ha **(9.3%)**
Covered soils	25,207 ha	23,314 ha	20,497 ha **(81.3%)**	26,371 ha	23,099 ha **(91.6%)**
Total	28,816 ha	28,816 ha	21,289 ha **(73.8%)**	28,816 ha	23,437 ha **(81.3%)**

Table 18.3. *Results of the land cover prediction with DST and DSmT*

18.3.2. *Integration of the climate variable in the land cover prediction model*

Space-time variability analysis of precipitation on the Scorff watershed pointed out that the climate variable was statistically significant for the land cover management in winter when precipitation was significant and characterized by a high spatial variability (see section 18.3.1). On the other hand, when its space variability was weak, the climate impact on the land use was not statistically significant. The integration of the climate variable (precipitation) in the land cover model was thus realized according to the first hypothesis.

18.3.2.1. *Mass function affectation of the climatic factor*

The mass function affectation for the climatic factor is derived from the analysis of the relationships between space-time variability precipitation and land cover for three winters (Table 18.4). For the hypothesis of a relative rainy winter characterized by high precipitation spatial variations, class 1 (< 100 meters) is concerned with a high probability to be in bare soils. On the contrary, for the parcels located in class 3 (> 170 meters), the probability to be in "covered soils" is clearly superior to the other "bare soils" hypothesis.

The five changes factors ("bare soils occurrence", "field size", "main crop precedent", "land cover evolution" and "climatic factor") are then combined with the two data fusion rules (DST and DSmT) according the same procedure (see section 18.3.1.2).

	Hypothesis "bare soils"	Hypothesis "covered soils"	Union of the hypotheses (DST)	Intersection of the hypotheses (DSmT)
Class 1 (<100 meters)	0.7	0.2	0.1	0
	0.7	0.2	0	0.1
Class 2 (100 - 170 meters)	0.3	0.5	0.2	0
	0.3	0.5	0	0.2
Class 3 (>170 meters)	0.1	0.6	0.3	0
	0.1	0.6	0	0.3

Table 18.4. *Mass function affectation for the climatic (precipitation) factor*

18.3.2.2. *Results of the land cover prediction with the integration of the climate variable*

The results of the land cover prediction (winter of 2003-2004) for the Scorff watershed with the climate factor integrated in the modeling process produced better final results since 82% (DST) and 83% (DSmT) of the parcels are correctly spatially predicted (Table 18.5). On the other hand, the prediction of the "bare soils" hypothesis was less precise than the previous land cover modeling that used only four information sources. This hypothesis still appears very difficult to spatially identify, even if the generated prediction was quite interesting with the two theories (2,648 ha for DST and 2,261 ha for DSmT, 3,609 ha detected with remote sensing). The spatial localization of the land use change for the "bare soils" was indeed very difficult to manage and the climate factor did not seem convincing to perform results. Furthermore, as we can see in Figure 18.4, the pattern fields were strongly divided, thus making the geographical location complicated. In addition, the winter of 2003-2004 was not especially rainy and that minimized the weight of the climate variable in the results. For the hypothesis "covered soils", which was mainly represented on the watershed, prediction and its spatial confrontation gave very interesting results with 92% of correct predictions for DST and 93% for DSmT.

	Land use in winter 2003/04	**Dempster-Shafer Theory**		**Dezert-Smarandache Theory**	
	remote sensing data (validation)	**Prediction**	**Right prediction**	**Prediction**	**Right prediction**
Bare soils	3,609 ha	2,648 ha	387 ha **(11%)**	2,261 ha	362 ha **(10%)**
Covered soils	25,207 ha	26,168 ha	23,182 ha **(92%)**	26,555 ha	23,544 ha **(93%)**
Total	28,816 ha	28,816 ha	23,569 ha **(82%)**	28,816 ha	23,906 ha **(83%)**

Table 18.5. *Results of the land cover modeling with the integration of the climate factor*

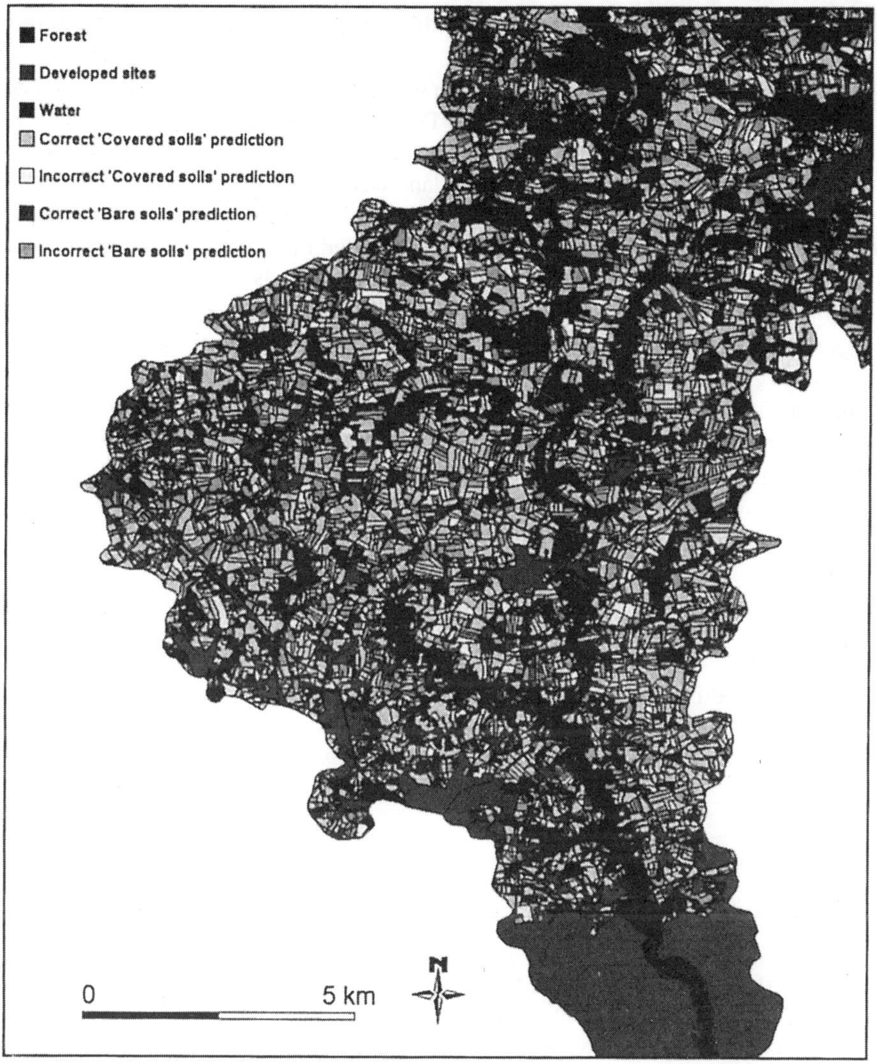

Figure 18.5. *Land cover prediction for the winter of 2003-2004 (south of the Scorff watershed)*

18.4. Conclusion

The land cover in winter in Brittany represents an important environmental task. Indeed, it corresponds to the rainiest season in Brittany (Western France) and the lack of vegetation on the agricultural fields can accelerate the flux transfers on the water resources. It can thus have a significant influence on the concentration of polluting matters (nitrogen, for example). On the Scorff watershed, a previous study tried to model the land cover evolution in winter by using a probabilistic model and data fusion rules (Dempster-Shafer and Dezert-Smarandache Theory). In this first study, the climatic factor was not considered, the pluviometry parameter is here taken in account and integrated into the land cover prediction modeling process. A previous analysis showed that precipitation is strongly linked to topography. Therefore, the climatic factor is indirectly integrated through a DEM. The statistical confrontation of the climatic factor with three winters shows that the pluviometry factor becomes significant on the land cover management in winter when the high precipitation amounts are characterized by high spatial variability between the North (altitude: 200 meters) and the coast (altitude: 40 meters). The integration of the climatic factor in the land cover prediction modeling process gives better results than the previous study but the results for the hypothesis "bare soils" characterized by high spatio-temporal changes and numerous driven factors are still insufficient. However, climate factor (precipitation) appears useful for the land cover evolution modeling: it enables a better understanding of this difficult and current problem.

18.5. Acknowledgements

This work is financed by the "Syndicat du Scorff" (Pont-Scorff, Morbihan) involved in the land cover monitoring with remote sensing data. We especially thank Thierry Mounier and Stéphanie Harrault for their valuable help in these researches.

18.6. Bibliography

[BAR 87] BARRY R.G., CHORLEY R.J., 1987: Atmosphere, weather and climate. London: Methuen.

[COR 02] CORGNE S., HUBERT-MOY L., BARBIER J., MERCIER G., SOLAIMAN B., 2002: Follow-up and modelling of the land use in an intensive agricultural watershed in France. 9th International Symposium on Remote Sensing, Greece, 23-27 September 2002, Proceedings of SPIE, 4879: Remote Sensing for Agriculture, Ecosystems, and Hydrology IV, pp. 342-351.

[COR 03] CORGNE S., HUBERT-MOY L., DEZERT J., MERCIER G., 2003: Land cover change prediction with a new theory of plausible and paradoxical reasoning. Fusion 2003 Conference, Cairns, Australia, 8-11 July, pp. 1141-1148.

[DEZ 03] DEZERT J., 2003, "Fondations pour une nouvelle théorie du raisonnement plausible et paradoxal", ONERA Tech. Rep. RT 1/06769/DTIM.

[HEG 00] LE HÉGARAT-MASCLE S., QUESNEY A., VIDAL-MADJAR D., TACONEY O., NORMAND M. and LOUMAGNE C., 2000: Land cover discrimination from multitemporal ERS images and multispectral Landsat images: a study case in an agricultural area in France, International Journal of Remote Sensing, vol. 21, no. 3, pp. 435-456.

[HEG 06] LE HEGARAT-MASCLE S., SELTZ R., HUBERT-MOY L., CORGNE S., NSTACH N., 2006: Performance of change detection using remote sensed data and evidential fusion comparison of three cases of application, International Journal of Remote Sensing, vol. 27, no. 15-16, pp. 3515-3533.

[KUL 59] KULLBACK S., 1959: Information theory and statistics, John Wiley and Sons, New York.

[LAM 94] LAMBIN E.F., STRAHLER A.H., 1994: "Indicators of land cover change for change vectors analysis in multitemporal space at coarse spatial scale", International Journal of Remote Sensing, vol. 15, no. 10, pp. 2099-2119.

[LED 01] LEDUC F., SOLAIMAN B. and CAVAYAS F., 2001: Combination of fuzzy sets and Dempster-Shafer theories in forest map updating using multispectral data, Sensor Fusion, SPIE, vol. 4385, pp. 323-334, Orlando, USA, 18-20 April, 2001.

[LEJ 83] LEJENÄS S.L. and ØKLAND H., 1983: Characteristics of Northern Hemisphere blocking as determined from a long time series of observational data, Tellus, 35, 350-362.

[MER 01] MERTIKAS P. and ZERVAKIS M.E., 2001: Exemplifying the Theory of Evidence in Remote Sensing Image Classification, International Journal of Remote Sensing, vol. 22, no. 6, pp. 1081-1095.

[MOR 90] MORON V., 1990: Contribution à l'étude de la variabilité climatique mensuelle en Europe, (unpublished) Thesis, Burgundy University, Dijon.

[PLA 05] PLANCHON O., 2005: Les printemps pluvieux des années 1980: retour sur une particularité météorologique et climatique de la fin du vingtième siècle en Europe occidentale. Actes du colloque de l'Association Internationale de Climatologie, Genes, 349-352.

[SHA 49] SHANON C.E. and WEAVER W., 1949: "The mathematical theory of communication", Chicago, University of Illinois Press Ill.

[SHA 76] SHAFER G., 1976: "A Mathematical Theory of Evidence", Princeton University Press, Princeton, New Jersey.

[SMA] SMARANDACHE F., DEZERT J.: Advances and Applications of DSmT for information Fusion. From Evidence to Plausible and Paradoxical Reasoning for Land Cover Change Prediction, American Research Press, Rehoboth, pp. 371-382, http://www.gallup.unm.edu/~smarandache/DSmT.htm.

[SRI 90] SRINIVASAN A. and RICHARDS J.A., 1990: Knowledge-based techniques for multi-source classification, International Journal of Remote Sensing, vol. 11, no. 3, pp. 505-525.

Chapter 19

A Tool for the Integrated Use of Remote Sensing with Ground Truth Data: DEMETER Project

19.1. Introduction

The needs and requirements of meteorological datasets for agriculture are related to the capability to integrate that data by taking into account the crops' spectral answer on the basis of the spatio-temporal relationships between weather impact and irrigation.

Irrigation Advisory Services (IAS) are natural management instruments to achieve a better efficiency in the use of water for irrigation. IAS are tools expected to help farmers to use water according to the crop water requirements.

Current IAS are labor and cost-intensive, yet they are unable to cover each field in extended areas at regular short time intervals. Earth Observation (EO) is naturally oriented to fit such a gap.

The new trends, driving forces and pressures related to water use for agriculture, which is assumed to be one of the most important users, highlight the increased

---

Chapter written by A. PERDIGAO, A. JOCHUM, A. CALERA, L. PESSANHA, A. CHINITA and J. MAIA.

importance concerning the knowledge of crop water requirements in every phenological stage, owing to an eco-compatible water use for agriculture.

The new capabilities and interfaces, data formats, structure and information at different levels of spatial aggregation highlight the importance of the developments that have been carried out within the frame of the DEMETER (Demonstration of Earth Observation Technologies in Routine Irrigation Advisory Services) project [CAL 99], and shows the importance between event and process. This project has been co-funded by the European Community (Energy, Environment and Sustainable Development Program, contract EVG1-CT-2002-00078) and has been simultaneously implemented in pilot zones in Spain, Italy, Portugal and Greece.

Modern systems provide a very high temporal and detailed amount of information, sometimes with very complex patterns and behaviors, which requires the development of tools with the capability to integrate a very dense and complex information process. This identifies the anomalies related to farmers and farmer's data and evaluates the quantity and quality of the information provided by each actor of the system.

The objective of the DEMETER project was to develop and demonstrate innovative solutions, while establishing mechanisms of permanent dialog between scientific-technical system developers and representative users.

19.2. Methodology used on the project

The current methodology used by IAS for this purpose is generally based on the standards recommended by FAO [ALL 98]. This method, which is also called the crop coefficient approach, consists of two parts. Firstly, reference evapotranspiration (Eto) is derived from measurements obtained from the IAS agrometeorological station. Secondly, the crop phenological phase is obtained across the IAS area. Potential evapotranspiration is then obtained from the multiplication of these two quantities. In the final step, crop water requirements are calculated as the difference between the observed precipitation and the potential evapotranspiration [ALL 98]. The irrigation scheduling information is then transferred to the end-user, i.e. the farmer, in various ways [TEI 01].

IAS must be set up according to the user needs by taking into account crop varieties and phenological stages. Crop water requirements models have been calibrated by using ground truth data from the zone covered, on the basis of soil crop relationship, so that they fit in with farming and irrigation parameters.

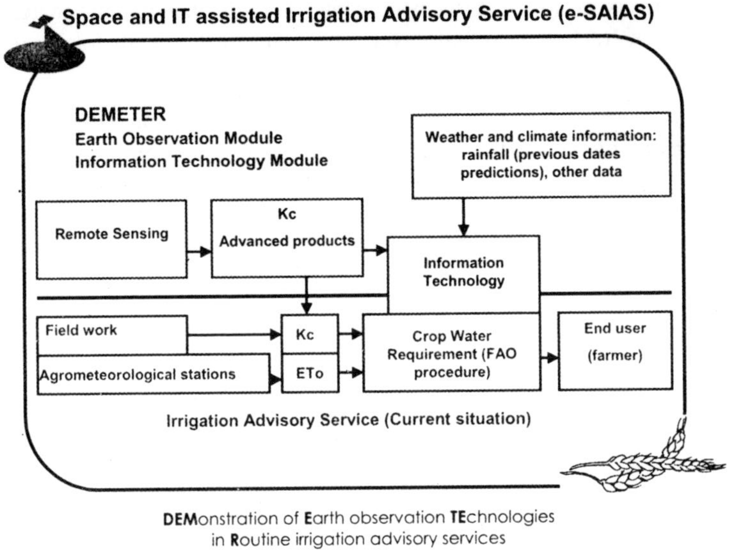

Figure 19.1. *Concept and functioning of current IAS (lower part) and new online Space-assisted Irrigation Advisory Service (eSaIAS)*

Besides the traditional interface based on a web application provided by the DEMETER project, the information technology (IT) currently available makes it possible to reach farmers in a more personal way and in near real-time. By using third generation mobile phones it is possible to deliver graphical and raster data of crops, plots and farm for each farmer. The platform design developed by the DEMETER project makes it possible, within 3 days of the image acquisition, to use EO-derived data and to deliver Kc spatial maps in each plot and the correspondent crop water requirements. The system also makes it possible to provide end-users with irrigation water supplied in each plot, thus updating in real-time the information supplied by the advanced IAS (see Figure 19.1).

19.3. Product line methodology

Ground truth campaigns have been carried out in each pilot zone to evaluate Kc Maps and advanced products for EO satellite data. They will also validate the system design at various levels and in its different components involving two levels of approach (see Figures 19.2 and 19.3):

– The *extensive campaign*, designed to collect the basic ground truth data which is for the estimation of the DEMETER "basic products" (i.e. Kc maps), is conducted in all the pilot zones covering measurements of crop phenological state, canopy height and green fractional ground cover, according to the methodology define by [ALL 98, CAL 01].

– The *intensive campaign* has been predicted to determine the relationships between crop the biophysical parameter and EO data under optimum crop growing conditions.

Some operational products for IAS derived from EO data have been obtained:

– Crop coefficient Kc by analytical approach.

– Ground Fractional Cover.

– Leaf Area Index (LAI).

– Accumulated (dry) Biomass.

And they make it possible to obtain additional information related to potential products like:

– Water stress indicator (water index).

– Evaporative fraction.

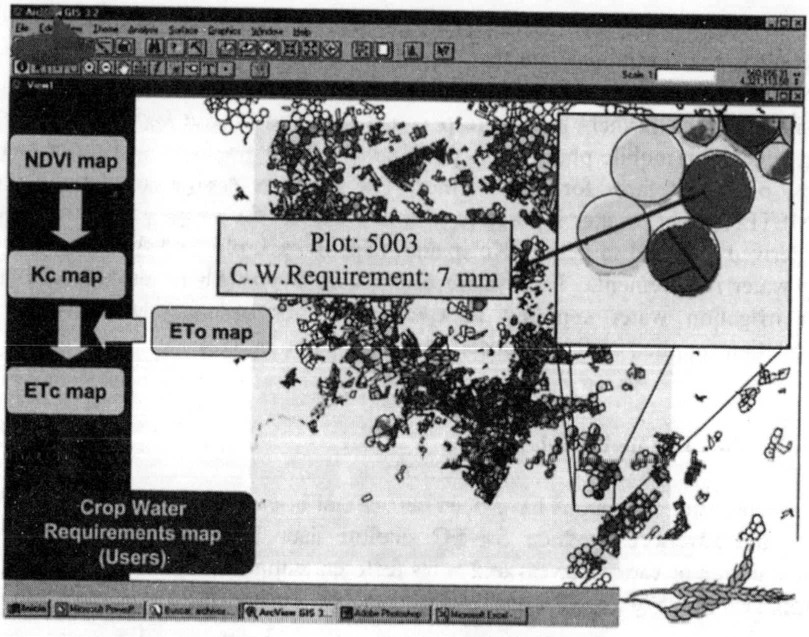

Figure 19.2. *Diagram for crop water requirements derived from the EO data source*

In the research pilot campaign the annual crops have been grown with farmer technology in order to establish the best development conditions, as near as possible to crop standard conditions (disease-free, well-fertilized, growing in large a field under optimum soil water condition and enabling full production under the given climatic condition).

The data processing consisted of the following steps:

– To provide spatial reference for each test plot by establishing a link with the GIS set-up for the Caia Irrigation District, as for the other pilot areas in Spain, Italy and Greece.

– To provide a crop classification inventory of the entire area, where a crop coefficient has been attributed to each plot, for the day when the measurement has been carried out.

– To apply the same procedure to the interpolated time series for each plot and for each day (indirect ground truth) in order to produce Kc maps, and supply information used to validate maps which are obtained from EO data.

The operational space segment was demonstrated in the three pilot zones. It was based on weekly Landsat 5 images (due to scene overlap) in Spain, alternating Landsat 5 and Landsat 7 in Portugal (due to the location of the pilot zone in the center of the L7 scene, which is not affected by the sensor failure) and a combination of Landsat 5, Ikonos, Spot and Aster in Italy (satisfying also the requirement for higher spatial resolution). The DEMETER basic products were generated and transmitted to a sample of farmers normally within 2 days from new coverage, thus completely matching the weekly operational irrigation scheduling cycles.

One of the major limiting factors for the use of EO technology in IAS has been the inadequate temporal resolution due to the long repeat cycle of adequate EO satellites. By using the full set of currently available high-resolution satellites (ASTER, IRS-1C, IRS-1D, Landsat 5-7 and SPOT 1-4, Quickbird and Ikonos) and low resolution satellites (MERIS, MODIS, Aqua-Terra, MSG and METOP, Vegetation) this problem will be overcome. High and very high spatial resolution time series will enable the overlook, in an acceptable time period, of crop development.

The potential use of data from recent innovative EO platforms in IAS, such as ASTER, (Advanced Spaceborn Thermal Emission and Reflection Radiometer) flying on Terra, provide a new perspective to the development of new management tools, like the prototype of the DEMETER project. ASTER provides important data, which is based on the characteristics of the available bands, like the VNIR, with the possibility to display images of the Earth in 14 different wavelengths of the electromagnetic spectrum, these range from visible to thermal infrared light in the SWIR (Shortwave infrared) at 30 m, basically because all three ASTER telescopes (VNIR, SWIR and TIR) are pointable in the cross-track direction. Its high resolution and its ability to change viewing angles in the TIR at 90 m, open the possibility to obtain significant new potentialities for its use, especially for agricultural purpose.

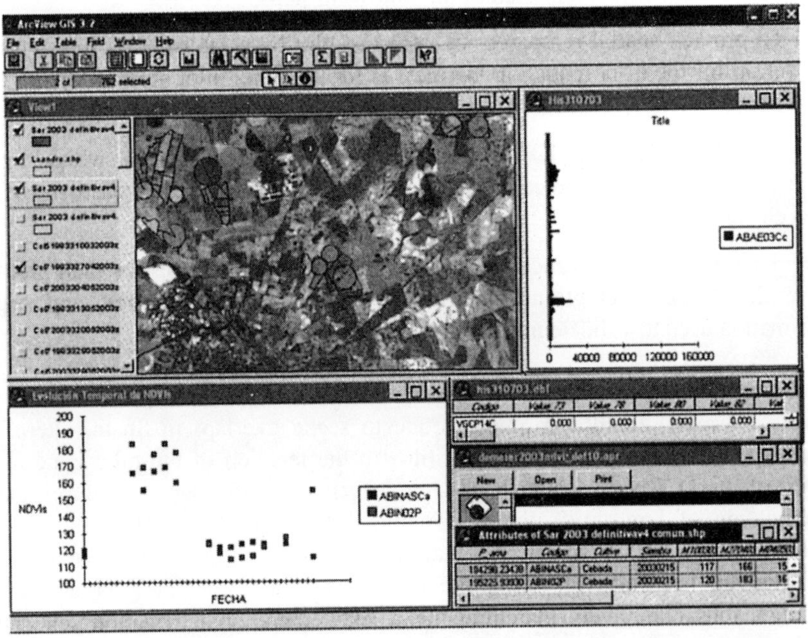

Figure 19.3. *Methodology for Kc calculation using GIS and image processing technologies*

A methodology for intersatellite calibration plays a fundamental role in solving the lack of temporal resolution, thus opening the way to obtain observations from different sensors which will be used jointly for monitoring crops throughout the growing season.

19.4. Report and data from the field campaigns

As in other zones, on the Caia pilot zone the campaigns covered the monitoring of the phenological state, crop height, fractional green cover, biomass, plant water content and LAI. This covers data that will be related to the calibration of Kc derived data (NDVI) and Kc (analytical expression). These two parameters have been derived from the ground truth data and the digital processing of satellite data time series, which cover the relationships towards the area covered (i.e. growing period, irrigation. practices and climatic conditions) owing to the analysis and set-up of Kc maps. A weekly procedure has been maintained in order to collect phenological stages data to define Kcb field curves (see Figure 19.3).

The basic product obtained in DEMETER from a sequence of (e.g. weekly) satellite images is maps of the crop coefficient Kc along with its temporal evolution at a given place. The crop coefficient is also the basic parameter used by traditional IAS for the estimation of crop evapotranspiration [ALL 98]. Therefore, the EO-derived Kc maps can be introduced directly into the routine IAS information generation flow for irrigation scheduling. Crossing (multiplying pixel by pixel) the Kc map with a map of reference evapotranspiration Eto (obtained from agrometeorological stations or by any other means) directly gives a map of crop evapotranspiration, which is then used to determine crop water requirements (see Figure 19.2).

Maps of crop water requirements have been derived from each image and introduced directly in the existing routine irrigation advisory operations. Here EO-derived information is expected to replace and complement data which was previously obtained by field work by using GIS to access the spatial dimension and enabling a more accurate prediction of water needs for irrigation (see Figures 19.4 and 19.5).

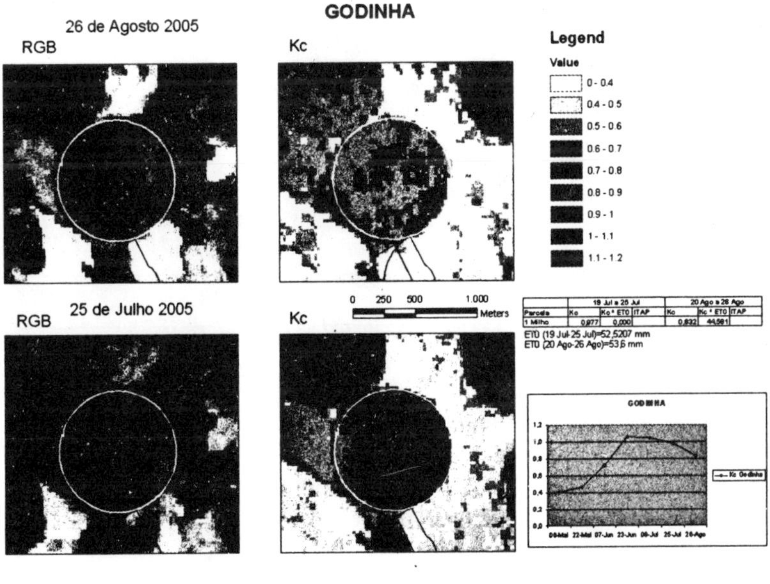

Figure 19.4. *Kc farmer report during the 2005 campaign in the Caia pilot area*

The major improvement achieved by the use of EO in the generation of basic IAS information products like crop coefficients is twofold. Firstly, the spatial coverage is geographical information system enhanced, both by extending to larger areas and by providing within-field heterogenity information. Secondly, the spatially resolved EO data can easily be combined with cadastral information in a geographical information system (GIS), which enables the personalization of the irrigation scheduling recommendation.

Spatial distribution of remotely sensed surface reflectance measurements to obtain canopy level structural and biophysical characteristics must be as precise as possible. This means that the transferring and biophysical property relationships must be at leaf level, where leaf optics interact with the canopy structure.

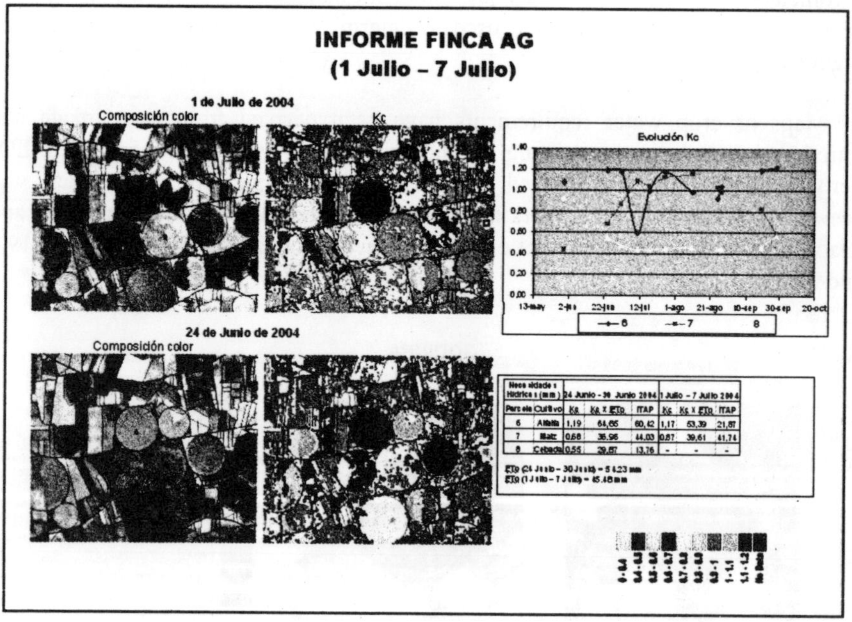

Figure 19.5. *Example of farmer's report distributed weekly to farmers' test groups in Spain*

Besides all the important issues of the project, field work is still considered necessary in order to provide regular ground truth for the online and off-line validation of the space-derived products, thus showing that the space segment is the most vulnerable part of the entire operational system.

19.5. Conclusion

IAS are the natural management instruments to achieve a better efficiency in the use of water for irrigation, helping farmers to apply water according to the actual crop water requirement and thus to maximize production and cost-effectiveness.

The operational demonstration consists of introducing EO-derived Kc maps into the IAS operational procedures. To assure the liability of crop coefficient (Kc) values based on EO data, the field experience and knowledge of the IAS manager played an important role in the evaluation between spatial data, crop water requirements and their phenological stage, as well for all the predicted crop-related information.

DEMETER also demonstrated that EO-derived data, when compared with a field survey, can be very useful to farmers and lead them to use IAS as a support to irrigation and farming strategies at their holding level. For this reason the farmers' feedback has been the main key to the success of the project.

The participatory evaluation with selected farmers shows that the farmers' feedback is very positive, both on the information quality and on the added value of the spatial information (within-plot heterogenity and between-plot variations). The reliability and accuracy of the information has been confirmed by the comparison of different approaches to derive crop coefficients from EO and the validation with field data in all pilot zones.

19.6. Acknowledgements

We would like to thank all the partners and the Heads of the following Institutions and teams involved on the DEMETER project, IDR, UVEG, ITAP, MAPyA, UNINA, INEA, MESTOR, DXSele, IM, IDRHa, UTAD, CHIRON, ULP, UTH.

19.7. Bibliography

[ALL 98] ALLEN R.G., PEREIRA L.S., RAES D. and SMITH M., 1998: Crop evapotranspiration: Guidelines for computing crop water requirements, FAO Irrigation and Drainage Paper 56. Roma.

[BAA 95] BAARS E., BASTIAANSEN A.P.M. and MENENTI M., 1995: A user-oriented and quantifiable approach to irrigation design. Water Res. Manag. 9(2): 95-113.

[BAS 00] BASTIAANSSEN W.G.M., MOLDEN D.J. and MAKIN I.W., 2000: Remote sensing for irrigated agriculture: examples from research and possible applications, *Agricultural Water Management*, 46, 137-155.

[CAL 99] CALERA A., MEDRANO J., VELA A. and CASTAÑO S., 1999: GIS tools applied to the sustainable management of water resources. Application to aquifer system 08-29, Agricultural Water Management, 40, 207-200.

[CAL 03] CALERA A., JOCHUM A.M. and CUESTA A., 2003: Space-assisted irrigation management: Towards user-friendly products. ICID Workshop on Remote Sensing of Crop Evaporation, Montpellier, 17 September 2003.

[CAL 01] CALERA A., MARTINEZ C. and MELIÁ J., 2001: A procedure for obtaining green plant cover. Its relation with NDVI in a case study for barley, International Journal of Remote Sensing, 22(7), 3357-3362.

[MAR 99] MARTÍN DE SANTA OLALLA F., BRASA RAMOS A., FABEIRO CORTÉS C., FERNABDÉZ D. and LÓPEZ CÓRCOLES H., 1999: Improvement of Irrigation management towards the sustainable use of groundwater in Castilla La Mancha, Spain, Agriculture Water Management, 40: 195-206.

[TEI 01] TEILLET P.M., BARKER B.L., MARKHAM B.L., IRISH R.R., FEDOSEJEVS G. and STOREY J.C., 2001: Radiometric cross-calibration of the Landsat-7 ETM+ and Landsat-5 TM sensors based on tandem data sets, *Remote Sensing of Environment,* 78, 39-54.

[URS 01] D'URSO G., 2001: *Simulation and management of on-demand irrigation systems*. PhD Thesis, Wageningen University, XI World Water Congress, Madrid, 5-9 October 2003.

[URS 96] D'URSO G. and MENENTI M., 1996: Performance indicators for the statistical evaluation of digital image classifications. ISPRS J. Photogramm. Rem. Sens. vol. 51 (2): 78-90.

Chapter 20

Assessing Population Exposure to Odorous Pollution from a Landfill over Complex Terrain

20.1. Introduction

Odors from landfill sites may lead to serious olfactory nuisances for communities that surround them. Though this pollution may be attributed to increases in the emission of landfill gases, they are usually associated with meteorological conditions, which drive their dispersion [SAR 03]. Landfill sites are often located over a complex topography for convenience which is mainly related to waste disposal and masking purpose. Consequently, the impacts of these odors (for instance in terms of population exposure) are difficult to assess [BOR 06]. To proceed, a geo-statistical approach is generally not possible because of scant measurements (odors cannot be easily measured) and difficulties in quantifying a subjective perception, which is variable in space and time, as well as in interpolating measurement points over a complex terrain. Conversely, numerical simulations might be a useful approach to access the details of small-scale atmospheric flow fields, which drive the dispersion of plumes of pollutants around a landfill site. Gaussian models are usually found to be insufficient to reproduce local atmospheric circulations since they are based on basic physics and oversimplified initialization. Nonetheless, we may note that the meteorological models are worth even less than the commonly used Gaussian models in accounting for the small-scale atmospheric processes if accurate input data is not available. The objectives of our study then are

Chapter written by Caroline RIESENMEY, Charles CHEMEL, Hervé VAILLANT and Mireille BATTON-HUBERT.

(*i*) to provide enough accurate terrain data to ensure a realistic dispersion of odors around a landfill site as calculated by a meteorological model and (*ii*) to use the model results to focus on assessing the population exposure to odorous pollution. For this purpose, GIS tools are applied and the accuracy of the data is discussed.

This chapter is organized as follows: in section 20.2, the landfill site and meteorological model are presented. Then, GIS tools are applied to supply ARPS with accurate terrain description of the landfill area. In section 20.3, selected model results are discussed and a map of the population exposure to odorous pollution is produced by using GIS tools. Finally, conclusions are given in section 20.4.

20.2. Model set-up

20.2.1. *Description of the landfill area*

The waste disposal facility is located over a hilly terrain near Saint-Etienne in the Rhône-Alpes-Auvergne French region (see Figure 20.1). The complex topography of this area makes the assessment of the airflow at least challenging. The landfill is one of the fifth largest landfills in France: the site receives about 500,000 tons of waste every year. The facility opened 30 years ago and should stop in about 15 years. It is situated in the vicinity of urban areas (including three cities with populations greater than 10,000 inhabitants). Thus, the landfill site is a potential source of odors, which may affect surrounding communities.

20.2.2. *Meteorological modeling*

Meteorological conditions leading to odorous pollution events have been identified by using a statistical analysis from meteorological data recorded *in situ* and an inventory of complaints from the surrounding communities [RIE 05]. Nine local weather conditions were identified and two of them were found to lead to the majority of the recorded complaints, that is, fair weather with high-pressure systems in both winter and summer. High-resolution numerical simulations of typical days which are representative of these two meteorological conditions have been conducted with the ARPS (Advanced Regional Prediction System) model [XUE 00]: a specific Eulerian model of dispersion was developed to trace odors and marked passive scalars were emitted from the work face every 15 minutes. Emission rates were calculated from concentration measurements realized during a working day. Several nested domains were used for the downscaling procedure (see Figure 20.2). The larger domain (Domain 1) was driven by the ECMWF (European Center for Medium-Range Weather Forecasts) gridded analyses. The finer grid (Domain 5) encompasses a domain of about 10 km x 10 km with a 100-m horizontal resolution.

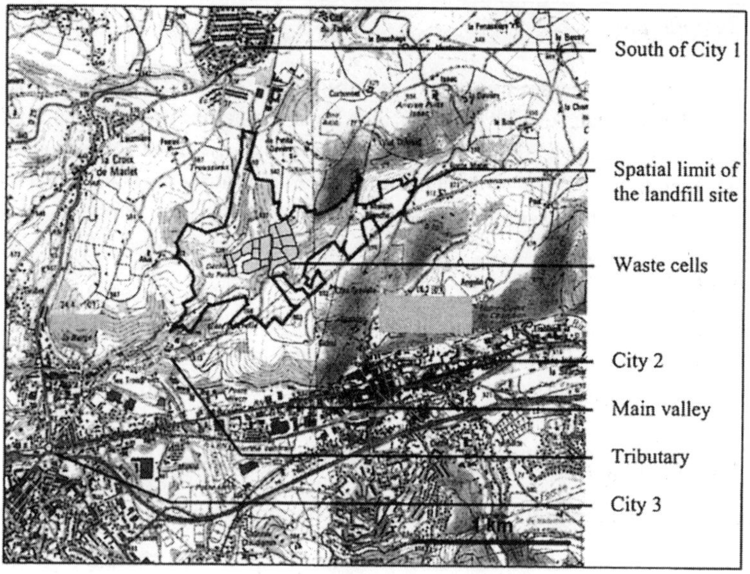

Figure 20.1. *Map of the landfill area*

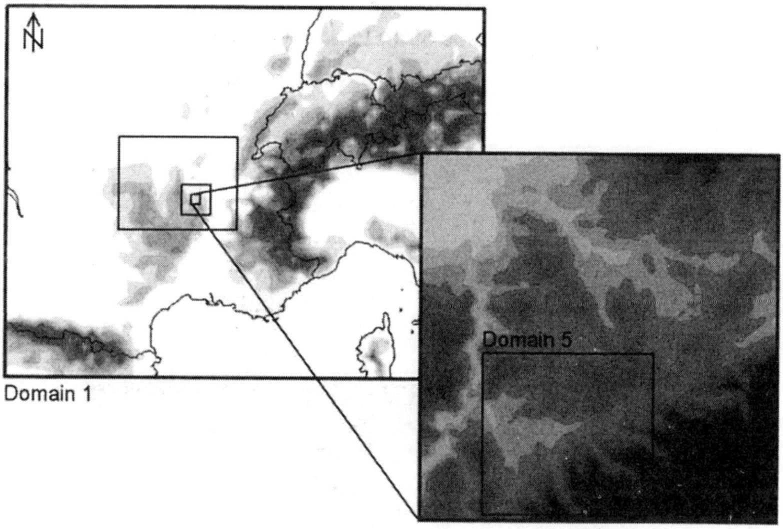

Figure 20.2. *Nested domains used for simulations with ARPS*

20.2.3. *Terrain data*

The terrain properties used in ARPS are orography, soil and vegetation types, and surface roughness. Data sets for Domain 1 to 3 were obtained from NGDC (National Geophysical Data Center). As for the terrain data for the two finer resolution domains (Domains 4 and 5), it was obtained from aerial photographs and IGN (National Geographic Institute) maps, and processed with the ArcGIS™ software.

20.2.3.1. *Orography*

Orography was handled from an IGN 1:125,000 scale topographic map covering the two finer domains (Domains 4 and 5). As an example, Figure 20.3 shows the six steps which were necessary to determine the altitude of each grid point in Domain 5, namely:

– georeferencing of the scanned IGN map in the Lambert II coordinate system;

– digitizing contours into a line layer and assigning an altitude for each line;

– linear interpolation of the elevation lines in a TIN (Triangular Irregular Network) data structure;

–raster encoding with a cell resolution of 10 m from the TIN layer;

– creation of raster with a cell resolution of 100 m from the existing 10-m layer. Each 100-m cell is assigned the value of the center 10-m cell;

– export of the attribute table in a text file containing the geographical coordinates and the elevation for each 100-m cell center.

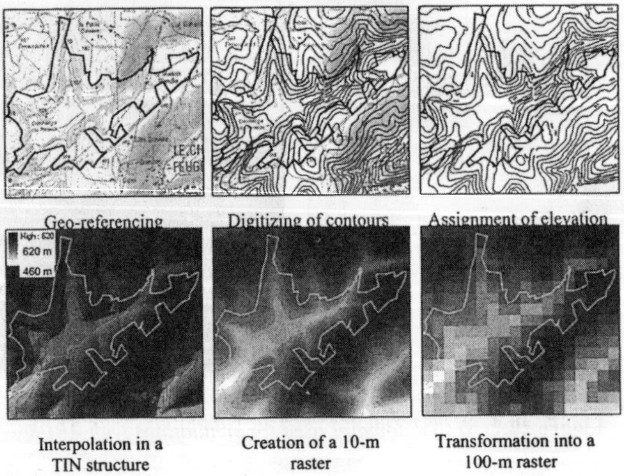

Figure 20.3. *Necessary steps to assign terrain elevation to the finer grid (Domain 5)*

Burying waste modifies the elevation, geometry and slopes of the landfill site. The difference between the starting and final altitudes can be up to 90 m in a waste dump. Thus, orography changes over time. These modifications are accurately assessed by land surveyors and the successive topographical maps are described in AutoCAD® format files. These files are imported into the ArcGIS™ software and used to create the contour defining the landfill area in ARPS (see Figure 20.4).

In order to obtain a raster which is representative of the hill and valley slopes, we need an intermediate step to convert the TIN layer into a 100-m raster. Creating a 10-m raster enables us to obtain the altitude at the center of the 100-m cell. Without this step, a 100-m raster from the TIN layer will only return the average elevation in each cell. The topography thus obtained is leveled out and summits and valleys are less prominent. Figure 20.5 shows the difference in the elevation with and without the 10-m raster creation step.

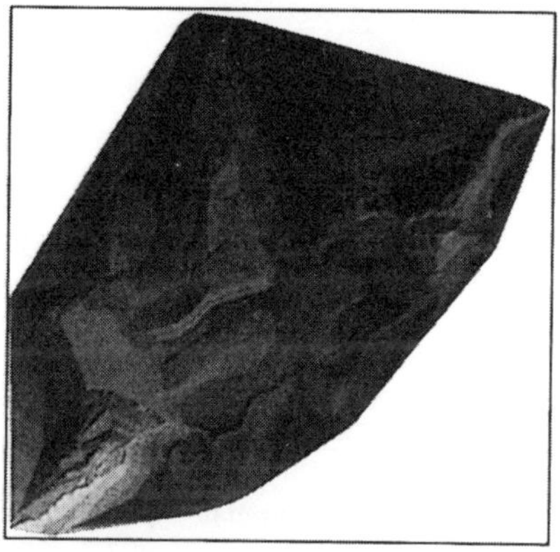

Figure 20.4. *TIN layer of terrain elevation obtained from the AutoCAD® file created by the land surveyor*

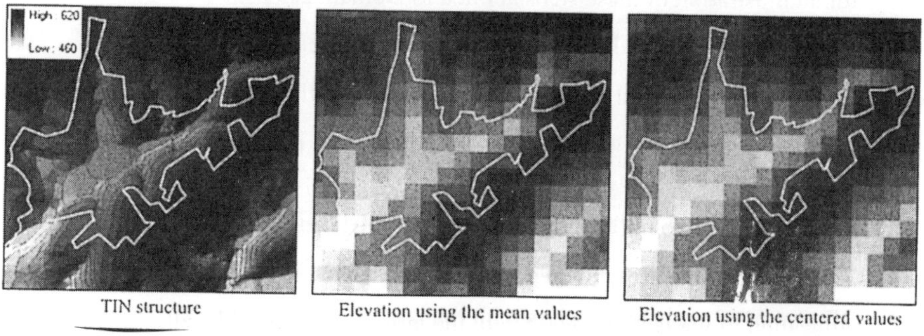

Figure 20.5. *Differences between mean and centered values for the elevation calculation*

20.2.3.2. *Land use*

Soil and ground surface properties were obtained from an IGN 1:125,000 scale topographical map and aerial photographs of the site. As for elevation, the ArcGIS^{TM} software was used to process the data. Firstly, 8 land use classes were identified as representative in Domains 4 and 5. Soil and vegetation types, as well as surface roughness were assigned to each class, according to Table 1.1.

N°	Land use	Vegetation type	Soil type	Surface roughness (m)
1	Road	Desert	Sand	0.002
2	City	Urban	Sand	0.01
3	No vegetation	Desert	Loam	0.002
4	Grassland	Grassland	Loam	0.01
5	Shrubs	Grassland with shrub cover	Loam	0.1
6	Forest	Deciduous forest	Loam	0.75
7	Waste	Desert	Clay loam	0.002
8	Water	Water	Water	0.001

Table 20.1. *Land use classes used for simulations with ARPS*

The necessary steps to integrate land use data in ARPS are as follows (see Figure 20.6 for an illustration of these steps applied to Domain 5):

– georeferencing of the IGN map and aerial photographs;

– digitizing the different land use classes by the creation of a polygon layer;

– creation of a second polygon layer for which each polygon represents a 100-m cell (creation from a 100-m raster);

–overlaying the two polygon layers and their intersections;

– calculation of the area covered by each land use class for each cell;

– allocation of the land use class with the largest area for each cell.

Contrary to orography, allocation of the centered value to the entire cell is not relevant. Furthermore, the use of centered values dramatically changes the results. Values obtained from the largest area in each cell are more representative and do not depend too much on the grid mesh.

Figure 20.7 summarizes the description of both orography and land use for Domain 5.

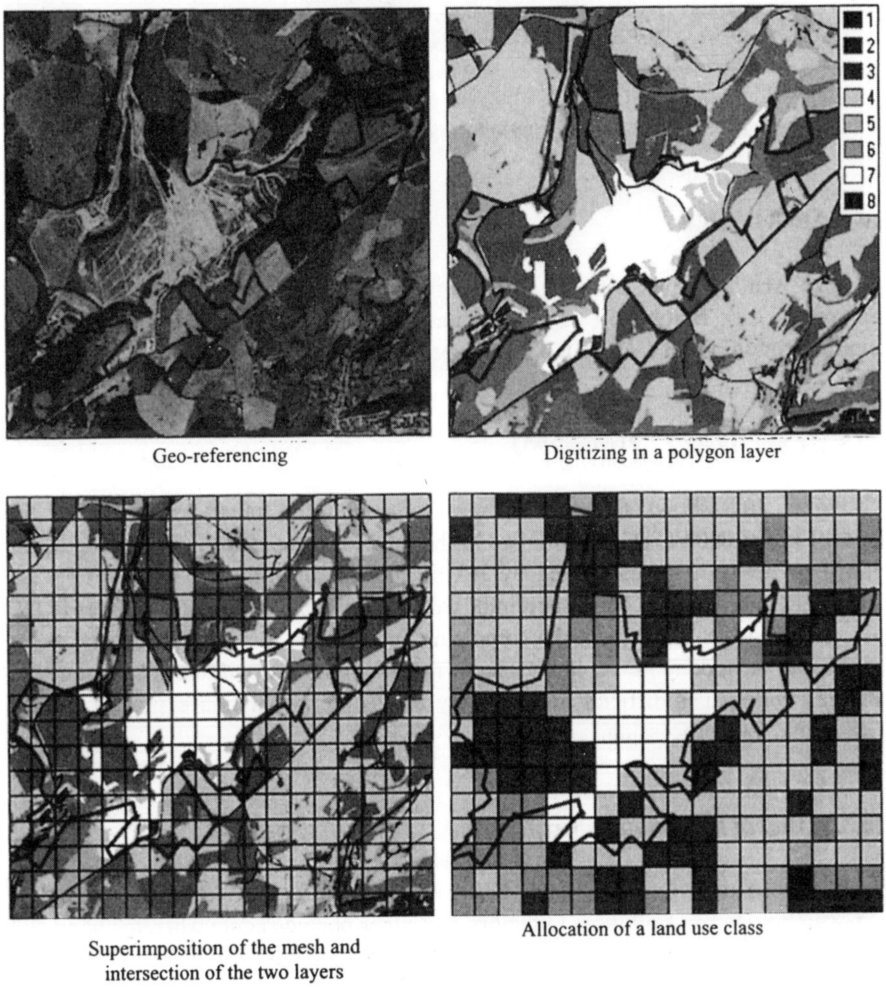

Figure 20.6. *Necessary steps to assign land use to the finer grid (Domain 5)*

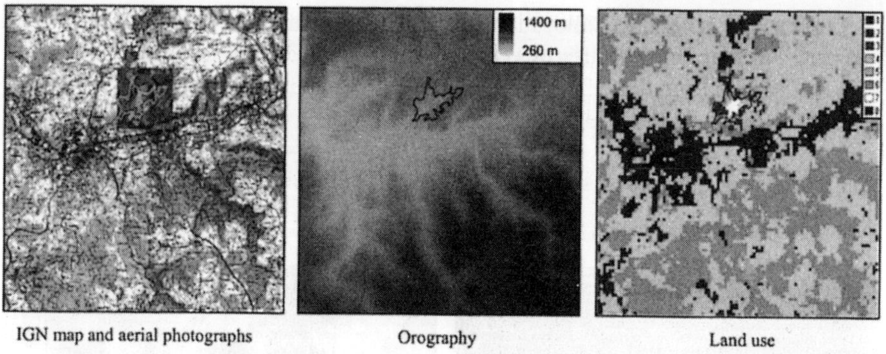

Figure 20.7. *Description of orography and land use for the finer grid (Domain 5)*

20.3. Model results

Two weather types were identified with a high risk of odorous pollution (see section 20.2.2). Below, the results obtained for a day which is representative of summer conditions (17 August 2002) are presented.

20.3.1. *Case study: 17 August 2002*

Figure 20.8 shows the dispersion of toluene (odor tracer) emitted from the landfill site at 0900 UTC (1100 LT). On the left is represented the dispersion of the plume of pollutant as calculated by ARPS and on the right the same results are integrated into ArcGIS™ and superimposed on a layer containing urban areas [KOU 02]. This result could explain the complaints recorded in the city located northward of the site at midday. Even if the model results highlight pollutant dispersion and affected zones, it is difficult to accurately assess the population exposure without GIS output data processing.

20.3.2. *Map of the population exposure*

In this section, we show how GIS tools can be used to quantify population exposure to odorous pollution. Olfactory nuisances become acute when the odors are very offensive and/or are all-encompassing. The method used in order to obtain the most affected zones is composed of three steps:

– normalization of the concentration field to obtain values comprised between 0 and 1: $C_{norm} = C/C_{max}$, where C is the concentration at the first level above the ground surface and C_{max} is the maximum concentration for the entire area;

– normalization of the population density: we use the ratio of the urban sector to the total area (population density data with a 100-m resolution is not available);

– multiplication of the two layers providing the population exposure.

Figure 20.9 shows the three 100-m rasters representing the normalized concentration, the normalized urban area and the population exposure obtained from the first two layers. This result is in excellent agreement with the inventory of complaints for this weather type: complaints occur in the morning in City 2 (south-eastward of the site) and in City 1 (northward of the site) in the middle of the day. The maximum value obtained in the third raster corresponds to a high-urbanized area and a weak concentration when compared with the maximal concentration obtained on the site. Note that the value representing the population exposure is not normalized, so that it would be possible to compare layers for different times (if the same normalization is used for all the times). Thus, the next step would be to superimpose rasters created from model results at different hours and to evaluate the most affected zones for each weather type.

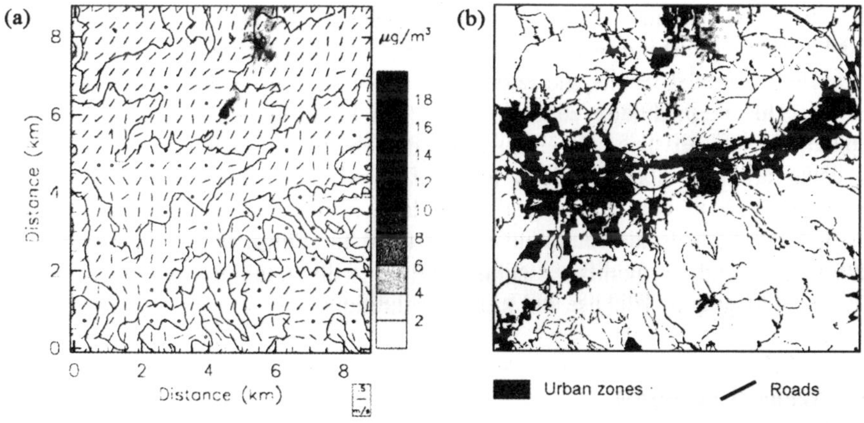

Figure 20.8. *(a) Toluene concentration field in the finer-resolved domain (Domain 5) at 0900 UTC (1100 LT) on 17 August 2002. (b) Same results integrated into ArcGIS™ and superimposed on a layer representing urban zones*

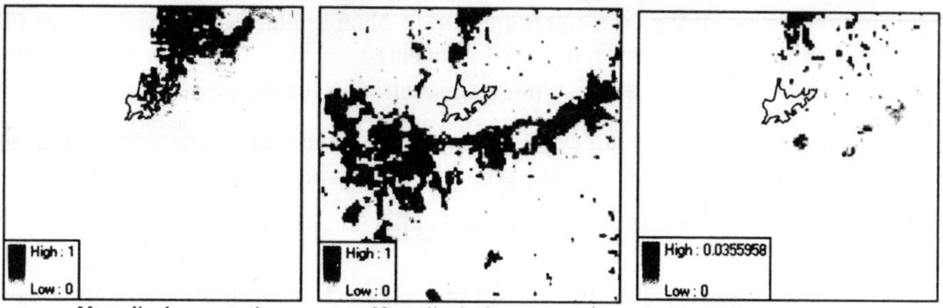

Normalized concentration — Normalized urban areas — Population exposure

Figure 20.9. *Assessment of population exposure to odorous pollution by combining pollutant dispersion with urbanization*

20.4. Conclusion

Deterministic modeling is a well adapted tool to represent meteorology and pollutant dispersion with a fine resolution over à complex terrain. It enables us to understand the local phenomena driving odor dispersion and to identify the affected zones. The use of GIS tools is necessary to provide input terrain data for such a modeling system (no database is available at a scale which is fine enough) and to analyze model outputs. The accurate assessment of population exposure is not possible without post processing the data. GIS was found especially useful to refine the excessive model results and correlate the pollution dispersion with population density. This methodology could be improved by using other data to describe population characteristics (e.g. sensitive people, children, presence of a hospital) or by using arbitrary normalizations in order to give a heavier weight to some parameters (use of non-linear functions). These results may be useful to provide decision-making aids for communities and firms to control the impacts of future landfill sites and to modify existing facilities with a tailored approach which considers topography, land use and weather conditions.

20.5. Acknowledgements

This research has been supported by the SATROD society, which is a subsidiary of the SITA Group, in charge of the activities on the landfill site. All major calculations were performed with IDRIS (Institut du Développement et des Ressources en Informatique Scientifique) and CINES (Center Informatique National de l'Enseignement Supérieur) computing resources. We are thankful to Chris Yukna for his help on editing this manuscript.

20.6. Bibliography

[BOR 06] BORJA-BAEZA R.-C., ESTEBAN-CHÁVEZ O., MARCOS-LÓPEZ J., PEÑA-GARNICA R.-J., ALCÁNTARA-AYALA I., "Slope instability on pyroclastic deposits: landslide distribution and risk mapping in Zacapoaxtla, Sierra Norte De Puebla, Mexico", *Journal of Mountain Science*, vol. 3 no. 1, 2006, p. 1-19.

[KOU 02] KOUSA A., KUKKONEN J., KARPPINEN A., AARNIO P., KOSKENTALO T., "A model for evaluating the population exposure to ambient air pollution in an urban area", *Atmospheric Environment*, vol. 36 no. 13, 2002, p. 2109-2119.

[RIE 05] RIESENMEY C., CHEMEL C., BATTON-HUBERT M., CHOLLET J.-P., "Mixed method to assess odor impact using data classification and high-resolution numerical simulations", *Proc. of the 5th Int. Conference on Urban Air Quality*, Valencia, Spain, 29-31 March 2005.

[SAR 03] SARKAR U., HOBBS S.-E., LONGHURST P., "Dispersion of odor: a case study with a municipal solid waste landfill site in North London, United Kingdom", *Journal of Environmental Management*, vol. 68 no. 2, 2003, p. 153-160.

[XUE 00] XUE M., DROEGEMEIER K.-K., WONG V., "The Advanced Regional Prediction System (ARPS) – A multi-scale non hydrostatic atmospheric simulation and prediction model. Part I: Model dynamics and verification", *Meteorology and Atmospheric Physics*, vol. 75 no. 3-4, 2000, p. 161-193.

Chapter 21

Disaggregated Estimation of N_2O Fluxes from Agricultural Soils of the Italian Region by Modelization in GIS Environment

21.1. Introduction

Nitrous oxide (N_2O) is one of the most important greenhouse gases, which, although present at a minor concentration in the atmosphere than CO_2, has a Global Warming Potential about 310 times greater than each molecule of CO_2 and a lifetime of about 120 years [IPC 00]. Its actual atmospheric concentration is about 314 ppbv (global burden 1,510 Tg N) and it is increasing at a rate of 0.25% per year [IPC 00]. Terrestrial ecosystems represent the main source of N_2O. About 60% of this source occurs in the Northern Hemisphere, where N_2O emissions are mostly associated with agricultural soils and arise from the fertilization with mineral N and animal manure, N derived from N_2-fixation (legumes and N-fixing microbes) and N from enhanced soil mineralization [DUX 93, MOS 97]. The indirect contribution of fertilizer N to N_2O emissions is also important. In fact, part of N applications will be lost as N_2O resulting from leaching, runoff, nitrogen oxides (NO_x) and NH_3 volatilization [MIN 90, DUX 93, MOS 97].

N_2O is a by-product of nitrification and an intermediate product of denitrification [GRA 94]. The process rate is controlled by environmental factors (soil carbon, quality and quantity of N inputs, temperature, soil texture, drainage, etc.) which

---

Chapter written by Anna CARFORA, Simona CASTALDI, Marco VIGLIOTTI and Riccardo VALENTINI.

influence the activity of these organisms, the denitrification being an anaerobic respiration process which represents the dominant source of N_2O in agroecosystems [GRA 94]. Territory characteristics which control losses of nitrogen from leaching, runoff, N gas volatilization and production (NO_x and NH_3) [BOW 02b, VAN 03], also represent distal controls of N_2O production.

A general and simple procedure to produce estimates of N_2O emissions from agricultural practices at national scale is proposed by the IPCC Guidelines [IPC 97]. In this technical document, the amount of direct emissions of N_2O from fertilizer addition as well as the amount of N losses via leaching or volatilization, which contribute to indirect N_2O production, are calculated by multiplying emission factors "EFs" by the rates of N applications. The environmental parameters which might influence N losses are not explicitly included in the calculations with few exceptions, where EFs might change with quality or semi-quantitative classes of environmental variables.

This approach presents many limitations. Environmental conditions vary widely between countries and within each country. In Italy, for example, climate variations from sub-humid to arid bioclimate, going from the Alps to the Southern regions, might significantly influence the amount of N which might be leached in different regions. Thus, in order to produce a more realistic estimate of N balance and losses such variations have to be included in the calculations in order to derive the overall N_2O budget. Also, N_2O emissions derived from the IPCC approach represent an aggregated value. On the other hand, a disaggregated approach helps to highlight critical areas for N losses and to understand the main driving factors (antrophic pressure, climate, soil characteristics, etc.) affecting N_2O emissions variations over time and space. This information might be particularly useful for management decisions.

In few cases, as an alternative approach to the IPCC methodology some studies have produced national estimates of N_2O emissions by using process-based biogeochemical models such as DNDC [LI 00] or Century [PAR 88]. These models are quite complex and require information at ecosystems level and relative to biological and abiotic soil parameters, with a level of detail which can be provided by few countries.

In this chapter a global estimate of the N_2O production from agricultural soils of the Italian regions was produced based on a spatial modelization of the most significant N losses and related N_2O fluxes, which are associated with agricultural practices. Such modelization was based on simple empirical models which relate N fluxes to heterogenous disaggregated georeferenced data of N inputs, agronomical information and key environmental characteristics.

21.2. Data sources and methods

21.2.1. *Methods applied for N_2O flux calculation*

In this procedure the agricultural N_2O emissions (E-N_2O) are calculated using the general approach proposed by the IPCC procedure where:

E-N_2O = **E-N_2O direct** (*fertilizers + histosols + crop residue + grazing + legumes*) + **E-N_2O indirect** (*NO and NH_3 deposition + N leaching*).

The direct emissions (E-N_2O direct) are emissions from agricultural fields due to anthropogenic N inputs. These are represented by synthetic fertilizers, animal excreta nitrogen (animal manure and excretal N deposited during grazing), N derived from biological N fixation, incorporation of crop residues left after harvesting into soils and soil nitrogen mineralization due to the cultivation of organic soils (histosols). The indirect emissions (E-N_2O indirect) are emissions derived from N deposition, nitrogen leaching and runoff.

21.2.2. *Applied models*

Direct N_2O emissions

In order to estimate the national emissions of N_2O deriving from the application of mineral fertilizers and animal manure, we used a simple model approach developed by [BOU 02b] and based on a dataset summarizing 846 emission measurements data collected from other works [BOU 02a]. The emission factor E (N_2O in kg ha^{-1} yr^{-1}) used to derive the amount of N_2O produced from the application of mineral N fertilizers and animal manure to crops and grasslands was equal to:

$$Emission = e^{\text{constant t} + \sum \text{Factor class}}$$

"Factor class" includes the factors with a significant effect on N_2O emission. These are environmental factors (climate, soil organic C content, soil texture, drainage and soil pH) and management-related factors (N application rate for fertilizer type, type of crop). Their weight is reported in [BOU 02b]. The derived emission factor is applied to the total amount of N deriving from the different N sources in order to get the N_2O production. N_2O emissions deriving from N fixed and N in crop residues were calculated by using [EME 04] default values. The N_2O deriving from this source was related to the land surface by crossing the data on land cover and crop distribution in the provinces.

Indirect N_2O emissions

N depositions were assumed to be approximately equal to the amount of reactive nitrogen, i.e. mainly NH_3-N and NO_x-N released in the area [EME 04].

Ammonia emission from synthetic fertilizers and animal manure applied to arable lands and grasslands were estimate by using a linear regression model derived for log-transformed weighted values of NH_3 volatilization rates as reported by [BOU 02c]. In their study the information taken from 1,667 NH_3 volatilization measurements documented in 148 research papers was summarized to assess the factors regulating the NH_3 loss from soil-plant systems to the atmosphere. Only factors that had a clear influence on NH_3 volatilization were used in the regression equation which was used to calculate the emission factor in parenthesis:

$$NH_3 \text{ volatilization rate} = e^{(\text{factor value\_crop type} + \text{fertilizer type} + \text{application mode} + \text{soil pH} + \text{soil CEC+climate})}$$

NH_3 volatilization losses from fertilizer and animal manure applications were calculated by multiplying the NH_3 volatilization rate for the total N input.

Similarly to NH_3 emissions, we used a REML-based model developed by [BOU 02b] to calculate the emissions of NO-N in kg ha^{-1} yr^{-1} which is derived from the application of mineral N fertilizers and animal manure to crops and grasslands. The factors included in the expression which account for NOx variability are N fertilizer type, soil organic C content and soil drainage.

N_2O emissions deriving from N deposition were calculated by using [EME 04] default values.

N_2O production derived from nitrate leaching was calculated firstly by estimating the leached N by using a model developed by [VAN 03] for soils under rainfed crops which combines the effect of temperature, crop type, soil properties and hydrological conditions on the annual mean nitrate leaching. In this model the annual leaching of nitrate from the root zone to the subsoil is calculated as:

$$N_{leached} = f_{leach} * N_{surplus}$$

$N_{surplus}$ is the annual surface N balance surplus, which is defined as the difference between the N inputs and N outputs for a given surface area of land. N inputs include biological N fixation, atmospheric N deposition, application of N fertilizers and animal manure and animal N excreted during grazing. N outputs in the surface N balance are NH_3 volatilization, N_2O and NO production, removal of N from the field by crop harvesting and grass consumption by animals. F_{leach} is the leaching fraction related to the environmental characteristics which are relevant for soil potential and which ranges from 0 to 1 [VAN 03] including precipitations, evapotranspiration, temperature, soil texture, drainage and soil organic matter.

Although the model has been produced for rainfed crops, in the calculation of the total input of water to the system also the water used for irrigation was taken into account and added to the mm of rain, where appropriate, as indicated by the land management. This slight modification was necessary as most of the data, which was used by [VAN 03] to produce the leached N fraction, derives from Northern European countries where rain contributes significantly to N leaching. In Italy, where, especially during the spring and summer crop growth, rain is very scarce, irrigation is a predominant source of water for N leaching.

For the calculation of N_2O deriving from leached N, the default value reported by [IPC 97] was used. No other relevant data has been published which proposes a different range of emission factors.

Data and software used for modelization

All the data used for the modelization were georeferenced. Calculations on the georeferenced data were done by using the map algebra procedure, which is a feature of the Idrisi Kilimanjaro GIS software (14.2), which runs on raster maps.

Input data

Agronomic data was derived from the database of the Italian National Institute of Statistic (ISTAT). The information was available at detail level of province. That data was georeferenced by distributing it homogenously over the province limits in the areas corresponding to the appropriate land use, as indicated by the CORINE land cover (2000 and 1990) maps (CORINE (Coordination of Information on the Environment 1:100,000 scale, UTM 32N on WGS84 datum and ellipsoid Coordinate Reference System). These categories included: arable land, rice fields, permanent crops (fruit trees, vineyard and olives groves), pastures and natural grassland.

Mean application rates of the N synthetic fertilizer for wetland rice fields, permanent crops and fertilized grassland were taken directly from [FAO 02]. The N fertilizer applications per hectare for upland crops were calculated as the difference between total country N fertilizer use [IST 00] and the sum of the N use in wetland rice fields, permanent crops and fertilized grassland divided by the area of upland crops. The mix of fertilizers types applied was assumed to be the same for grasslands and croplands: some of the fertilizer categories used are single N fertilizers (ammonium sulphate, ammonium nitrate, calcium ammonium nitrate, urea and other straight N) and others are compound fertilizers of two or more different fertilizers types (binary as NP and NK and ternary as NPK). It was also assumed that, on a general base, fertilizers were applied as a basal application by broadcasting.

The amount of nitrogen used as fertilizer from animal excreta was estimated by using the animal population data of the country for non-dairy and dairy cattle, pigs, poultry, horses, sheep and goats (5th Agricultural Census, 2000) in combination with default values on annual N excretion rates for each animal type and region [ECE 94, VET 89, STE 90] and fractions of manure N used in different animal waste management systems from [SAF 92]. The amount of animal manure available for application as a fertilizer was calculated as being equal to the total animal manure production, excluding the manure used as fuel and the manure excreted during grazing. Sheep and goats are essentially grazing animals and their manure was considered to be unavailable for spreading. Manure N application rates for upland crops and grasslands were assumed to be equal, and manure and N additions during grazing were assumed to be applied in pastures and natural grasslands.

Within the upland crops, we distinguished leguminous crops (pulses and soybeans) which fix atmospheric N and the biological fixation was calculated from the crop dry matter production data of pulses and soybeans (ISTAT). For the estimation of the atmospheric N fixed by N-fixing crops it was assumed that the amount of N contained in the above-ground plant material (crop product plus residues) is a reasonable proxy for the total amount of N fixed by the crop. The data of total leguminous crop production (pulses and soybeans) within upland crops was taken from [IST 00] and following the approach presented in the IPCC Guidelines the amount of fixed N was calculated by multiplying the seed yield transformed in dry matter assuming 15% water content, by a default value of 2 that converts the crop production to total crop biomass (assuming that the mass ratio of residue to product is 1) and then by the fraction of crop biomass that is nitrogen (0.03 kg N/kg of dry biomass).

The amount of N returned to the soils annually through the incorporation of crop residues was calculated from crop dry matter production data (ISTAT) by adjusting the amount of total above-ground crop biomass that is removed from the field as product and for the fraction that is burned.

Georeferenced environmental data used for the calculations included soil organic carbon content, soil drainage, pH, CEC (FAO-UNESCO Soil Map of the World, 2003, raster maps scale 1:5,000,000), soil texture (European Soil Database http://eusoils.jrc.it), rainfall intensity, evapotraspiration rate and air temperature (MARS-STAT Database, (ftp://agrifish.jrc.it/Public/CGMS/shape/grid50.zip).

In order to perform simulations, all the input data was first transformed so as to get the same geographic reference system (UTM 32N) and then processed so as to have the same spatial resolution of CORINE land cover data (25 hectares).

21.3. Results and discussion

The total emission of N_2O from agricultural soils in Italy was estimated to be 46.5 Gg N_2O yr^{-1} in 2000, having had an increment of about 15% from the previous decade (total emissions in 1990, 39.6 Gg N_2O yr^{-1}).

The disaggregated cumulative yearly N_2O emissions, calculated for the whole Italian agricultural land are reported in Figure 21.1c. In Figures 21.1a and 21.1b a detail is given of the estimated N_2O fluxes from the direct and indirect sources, for the year 2000. Big differences in N_2O fluxes can be observed along the Italian territory and this is the result of many factors. First of all, the big differences in N input from agricultural practices (Table 21.1) and second, the climatic and edaphic factors, which are particularly relevant for N leaching but also for NH_3 volatilization.

Overall, in 2000 the highest N_2O emissions were found in the Northern regions of Emilia-Romagna, Lombardia and Veneto (Figure 21.2) where there is a very high exploitation of land by intensive agriculture and a high concentration of animal husbandry (Table 21.1). Emissions are lower in the Southern regions (Figure 21.1c) with the exception of Puglia and Sicily (Figure 21.2), although there the agriculture, which is predominantly based on crops, is less intensive than in the North and has much lower investments in animal husbandry (Table 21.1).

Considering the total Italian agricultural surface N_2O direct and N_2O indirect emissions contributed for about 58 and 41%, respectively, to the total source strength (27.3 and 19.1 Gg N_2O yr^{-1}, respectively) in 2000 and for about 65 and 35%, respectively, (25.85 and 13.8 Gg N_2O yr^{-1}, respectively) in 1990. The lower contribution of indirect sources to N_2O production observed in 1990, compared to 2000, seems to depend mainly on the lower contribution of N_2O deriving from leached surplus N, which might be attribute to a lower excess of rainfall over evapotranspiration and only in a minor part to lower net inputs of N to the soil.

The application of mineral fertilizers and animal manure by using the model approach developed by [BOU 02b], yielded a derived annual N_2O emission which is lower than the rate reported by the Italian National Environmental Agency for the same year. This difference is the result of the different approach used to calculate N_2O fluxes. In fact, the IPCC procedure applies a fixed emission factor of $1.25 \pm 1.0\%$ to derive the fraction of fertilizer N which is lost as N_2O, whereas by applying [BOU 02b] model a lower emission factor is obtained and this factor changes over the Italian territory depending on the type of fertilizer applied, crop type, soil texture, soil organic matter content and soil pH.

Overall, the spatial disaggregated modeling approach we used made it possible to evidence big spatial differences over Italy, which can be attributed to differences in land use, environmental characteristics such as soil and climate and different agricultural management. The empirical equations, used to derive N fluxes, produced estimates which are quite close to the estimates which can be obtained by IPCC procedure, thus showing a certain robustness and accord in both approaches. In the calculation of indirect emissions, the present approach evidenced the importance of climatic conditions which significantly influence the amount of N leached from the systems which can be subsequently nitrified and denitrified in aquatic ecosystems with the production of N_2O. In fact, this source of N_2O was a lot less in the Southern regions which are characterized by a lower precipitation/ evapotranspiration ratio. Generally the amount of N lost as leachate is much higher than the fraction of N directly lost as gaseous products. This, as well as other spatial information, can be used to better control and plan the management options for future scenarios in agroecosystems.

Region	N Fert	Number of animals (thousands)				Nex	CropBF	Nfix	CropO	Ncr	**Total N input**	TS	AS*
	Gg N	Cattle	Pigs	Sheep	Poultry	Gg N	Gg d.m.	Gg N	Gg d.m.	Gg N	**Gg N**	**000 ha**	**000 ha**
Abruzzo	19.8	83	115	256	3,602	12.5	11.7	0.7	1143.5	15.8	**48.8**	1083.1	627.6
Basilicata	10.9	78	83	296	496	13.0	5.0	0.3	945.0	12.9	**37.1**	1007.2	624.0
Calabria	16.5	102	101	210	1,412	14.9	27.8	1.7	1048.6	14.9	**47.9**	1522.3	797.0
Campania	47.0	343	142	199	5,766	28.5	80.5	4.8	1869.5	27.4	**107.7**	1367.2	815.0
Emilia-Romagna	94.3	623	1,553	73	29,088	80.8	173.2	10.4	8040.5	113.2	**298.6**	2212.1	1522.1
Friuli	30.1	101	192	5	8,638	14.6	123.8	7.4	1289.3	20.7	**72.8**	786.1	338.6
Lazio	40.9	273	89	586	3,323	27.8	18.8	1.1	1747.1	24.1	**93.9**	1723.0	1025.7
Liguria	2.2	17	2	17	279	1.7	2.3	0.1	53.4	0.8	**4.8**	540.9	107.5
Lombardy	112.2	1,611	3,840	75	27,286	173.0	155.1	9.3	4914.9	70.5	**365.0**	2386.7	1231.8
Marche	40.2	78	148	146	7,693	13.2	27.1	1.6	2177.9	30.1	**85.2**	975.0	661.4
Molise	5.5	57	47	93	4,034	8.0	1.5	0.1	505.2	6.9	**20.5**	445.9	287.4
Piemont	63.8	819	924	75	13,967	70.5	102.8	6.2	3273.9	47.0	**187.5**	2538.3	1303.1
Puglia	60.2	158	27	200	1,982	15.3	28.8	1.7	5284.0	72.1	**149.4**	1953.9	1671.0
Sardinia	18.9	250	194	2,542	1,139	55.4	21.7	1.3	735.0	10.5	**86.1**	2399.2	1208.8
Sicily	53.9	308	42	617	1,678	29.6	71.3	4.3	1893.7	27.5	**115.3**	2572.8	1886.2
Tuscany	45.4	104	172	495	3,484	17.6	18.2	1.1	1411.9	19.6	**83.6**	2294.8	1056.9
Trentino	5.7	189	22	45	1,362	14.1	0.1	0.0	47.4	0.6	**20.4**	1359.8	316.6
Umbria	26.8	63	250	137	8,170	14.4	1.2	0.1	942.0	12.7	**54.0**	846.1	454.5
Aosta Valley	0.0	39	1	2	15	2.5	0.0	0.0	6.4	0.1	**2.6**	325.9	61.1
Veneto	104.2	933	702	27	47,983	91.8	338.7	20.3	5454.7	82.8	**299.1**	1842.4	1111.2
Total	798.4	6,230	8,645	6,096	171,399	699.1	1,209.4	72.6	42,783.9	610.2	**2,180.3**	30,182.7	17,107.4

Table 21.1. *Total mineral fertilizer use (N fert), number of animals (thousands), manure N excretion (Nex), total N-fixing crop yield (CropBFix), the amount of N fixed (Nfix), total other crop yield (CropO), N in crop residues(Ncr), total N input, total land surface (TS) and agricultural area (AS) reported for each region of Italy in 2000*

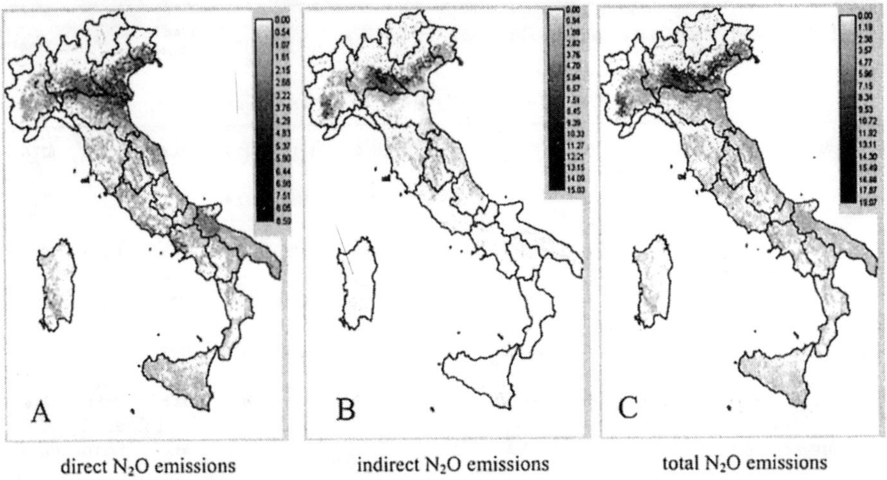

Figure 21.1. *Disaggregated N_2O emissions obtained from direct sources (A), indirect sources (B) and total emissions (C) calculated for the Italian agricultural territory in 2000. Units of measure represent kg N_2O ha^{-1} $year^{-1}$*

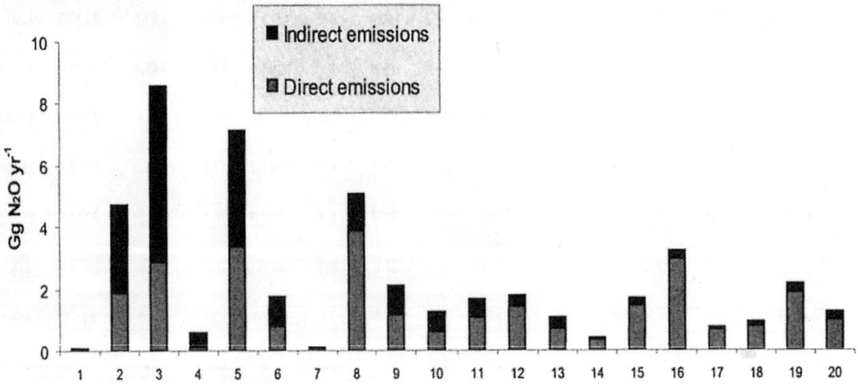

Figure 21.2. *Histograms representing the contribution of direct (grey bars) and indirect (black bars) sources to the total N_2O production for each Italian region. Regions are distributed from 1 to 20 following a climatic gradient North-South. 1: Aosta Valley; 2: Piedmont; 3: Lombardy; 4: Trentino-Alto Adige; 5: Veneto; 6: Friuli-Venezia Giulia; 7: Liguria; 8: Emilia-Romagna; 9: Tuscany; 10: Umbria; 11: Marche; 12: Lazio; 13: Abruzzo; 14: Molise; 15: Campania; 16: Puglia; 17: Basilicata; 18: Calabria; 19: Sicily; 20: Sardinia*

21.4. Bibliography

[BOW 02] BOUWMAN A.F., BOUMANS L.J.M. and BATJES N.H., (2002a). Emissions of N_2O and NO from fertilized fields: Summary of available measurement data. Global Biogeochemical Cycles, 16(4), 1058, doi:10.1029/2001GB001811.

[BOW 02b] BOUWMAN A.F., BOUMANS L.J.M. and BATJES N.H., (2002b). Modeling global annual N_2O and NO emissions from fertilized fields. Global Biogeochem. Cycles, 16(4), 1080, doi:10.1029/2001GB001812).

[BOW 02c] BOUWMAN A.F., BOUMANS L.J.M. and BATJES N.H., (2002c). Estimation of global NH3 volatilization loss from synthetic fertilizers and animal manure applied to arable lands and grasslands. Global Biogeochem. Cycles, 16(2), 1024, doi:10.1029/ 2000GB001389.

[DUX 93] DUXBURY J.M. and MOSIER A.R. (1993). Status and issues concerning agricultural emissions of greenhouse gases. In Agricultural dimension of the Global Climate Change, pp. 229-258. Kaiser H.M. and Drennen T.E. (Eds.), St Louis Press. Delray Beach, FL.

[DUX 93] DUXBURY J.M., HARPER L.A. and MOSIER A.R. (1993). Contributions of agroecosystems to global climate change. In Agricultural Ecosystems Effects on Trace Gases and Global Climate Change. pp.1-18. Harper A.R., Mosier J.M., Duxbury J.M. and Rolston D.E. (Eds.). ASA special Pub. 55, Am. Soc. Agron. Inc. Madison, WI.

[ECE 94] ECETOC (1994). Ammonia emissions to air in Western Europe. Technical Report No 62. Brussels.

[EME 04] EMEP/CORINAIR (2004). Emission Inventory Guidebook. European Environment Agency (EEA), Technical report No 30.

[FAO 02] Food and Agriculture Organization (FAO), International Fertilizer Industry Association (IFA) and International Fertilizer Development Center (IFDC) (2002). Fertilizer use by crop, 5th ed., Rome.

[GRA 94] GRANLI T. and BØCKMAN O.C. (1994). Nitrous oxide from agriculture. Norw. J. Agr. Sci., supp.12.

[IPC 00] IPCC (2000). Emissions Scenarios. 2000. Special Report on the International Panel on Climate Change (eds. N. Nakicenovic and R. Swart), pp. 570. Cambridge University Press, UK.

[IPC 97] IPCC (1997). Organization for Economic Cooperation and Development (OECD) and International Energy Agency (IEA). Revised 1996 Guidelines for National Greenhouse Gas Inventories, Chapter 4. Agriculture: Nitrous oxide from agricultural soils and manure management, OECD, Paris, France.

[ITA 00] Italian National Institute of Statistic (ISTAT). (2000). Annuario Statistico, 5th Agricultural Census.

[LI 00] LI. (2000). Modeling trace gas emissions from agricultural ecosystems. Nutrient Cycling in Agroecosystems 58:259-276.

[MIN 90] MINAMI K. and OHSAVA A. (1990). Emissions of nitrous oxide dissolved in drainage water from agricultural land. In Soil and the Greenhouse Effect, pp. 503-509. Bouwman A.F. (ed.), John Wiley and Sons, New York.

[MOS 93] MOSIER A.R. (1993). Nitrous oxide emissions from agricultural soils. In Methane and nitrous oxide: methods in national emission inventories and options for control proceedings, pp. 273-285. Van Amstel A.R. (ed.), National Institute of Public Health and Environmental Protection, Bilthoven, The Netherlands.

[MOS 97] MOSIER A.R., KROEZE C., NEVISON C., OENEMA O., SEITZINGER S. and VAN CLEEMPUT O. (1997). Closing the global atmospheric N_2O budget: nitrous oxide emissions through the agricultural nitrogen cycle. Nutr. Cycl. Agroecosystems, 52, pp. 225-248.

[NEV 96] NEVISON C.D., ESSER G. and HOLLAND E.A. (1996). A Global Model of Changing N_2O Emissions from Natural and Perturbed Soils. Climatic Change, 32, pp. 327-378.

[PAR 88] PARTON W.J., STEWART J.W.B. and COLE C.V. (1988). Dynamics of C, N, P and S in grassland soils: a model. Biogeochemistry 5:109-131. (472).

[SAF 92] SAFLEY L.M., CASADA M.E., WOODBURY J.W. and ROOS K.W. (1992). Global methane emissions from livestock and poultry manure. USEPA Report No. 400/1-91/048. Office of Air and Radiation, Washington DC.

[SMI 98] SMITH K.A., THOMSON P.E., CLAYTON H., MCTAGGART I.P. and CONEN F. (1998). Effects of temperature, water content and nitrogen fertilization on emissions of nitrous oxide by soils. Atmos. Environ., 32, pp. 3301-3309.

[STE 90] STEFFENS G. and VETTER H. (1990). Neue Faustzahlen ueber Nachrstoffgehalte und Naehrstoffanfall. Landwirtsschaftsblatt Weser-Ems, 3.

[VAN 03] VAN DRECHT G., BOUWMAN A.F., KNOOP J.M., BEUSEN A.H.W., MEINARDI C.R. (2003). Global modelling of the fate of nitrogen from point and nonpoint sources in soils, groundwater, and surface water. Global Biogeochem. Cycles, 17(4), 1115, doi:10.1029/2003GB002060.

[VET 89] VETTER H., KLASINK A. and STEFFENS G. (1989). Mist und Guelledue 1 ngung nach Mass. VDLUFA-Schriftenreihe, 19, pp. 41-66.

List of Authors

C. ALMARZA
Instituto Nacional de Meteorología
Spain

Carlos Frederico ANGELIS
Instituto Nacional de Pesquisas Espaciais – INPE
Centro de Previsão de Tempo e Estudos Climáticos – CPTEC
Brazil

Ingeborg AUER
ZAMG
Vienna, Austria

Malgorzata BARSZCZYNSKA
Institute of Meteorology and Water Management
Kràków, Poland

Mireille BATTON-HUBERT
ENS des Mines
Saint-Etienne, France

Szabolcs BELLA
Hungarian Meteorological Service
Hungary

Zita BIHARI
Hungarian Meteorological Service
Hungary

Reinhard BÖHM
ZAMG
Vienna, Austria

A. CALERA
IHERA, Ministry of Agriculture
Lisbon, Portugal

Anna CARFORA
Dipartimento di Scienze Ambientali
Università degli Studi di Napoli
Caserta, Italy

Simona CASTALDI
Dipartimento di Scienze Ambientali
Università degli Studi di Napoli
Caserta, Italy

Lee CHAPMAN
School of Geography, Earth and Environmental Sciences
University of Birmingham
UK

Charles CHEMEL
LEGI
Joseph Fourier University
Grenoble, France

A. CHINITA
IHERA, Ministry of Agriculture
Lisbon, Portugal

S. CORGNE
COSTEL Laboratory
University of Rennes 2
France

T. CORPETTI
COSTEL Laboratory
University of Rennes 2
France

Cintia Pereira DE FREITAS
Instituto Nacional de Pesquisas Espaciais – INPE
Centro de Previsão de Tempo e Estudos Climáticos – CPTEC
Brazil

J. DE LA CRUZ
Consejería de Educación
Comunidad Autónoma de Madrid
Spain

Pierre DUMOLARD
Laboratoire SEIGAD
Joseph Fourier University
Grenoble, France

Izabela DYRAS
Institute of Meteorology and Water Management
Kraków, Poland

A. JOCHUM
IHERA, Ministry of Agriculture
Lisbon, Portugal

Claude KERGOMARD
Ecole Normale Supérieure
Paris, France

Danuta KUBACKA
Institute of Meteorology and Water Management
Kraków, Poland

Rémi LHOTELLIER
Laboratoire SEIGAD
Joseph Fourier University
Grenoble, France

Jean-Christophe LOUBIER
Department of Geography
University of Geneva
Switzerland

M.Y. LUNA
Instituto Nacional de Meteorología
Spain

Pawel MADEJ
Institute of Meteorology and Water Management
Kraków, Poland

J. MAIA
IHERA, Ministry of Agriculture
Lisbon, Portugal

M.L. MARTIN
Dpto. Matemática Aplicada, Escuela Universitaria de Informática
Universidad de Valladolid
Segovia, Spain

Fabiano MORELLI
Instituto Nacional de Pesquisas Espaciais – INPE
Centro de Previsão de Tempo e Estudos Climáticos – CPTEC
Brazil

Ákos NÉMETH
Hungarian Meteorological Service
Hungary

Cristian Valeriu PATRICHE
Romanian Academy, Department of Iasi, IASI
Romania

Antonio PERDIGAO
IHERA, Ministry of Agriculture
Lisbon, Portugal

L. PESSANHA
IHERA, Ministry of Agriculture
Lisbon, Portugal

O. PLANCHON
COSTEL Laboratory
University of Rennes 2
France

H. QUÉNOL
COSTEL Laboratory
University of Rennes 2
France

Caroline RIESENMEY
ENS des Mines
Saint-Etienne, France

Wolfgang SCHÖNER
ZAMG
Vienna, Austria

Alvaro SILVA
Portuguese Meteorological Institute
Portugal

M.G. SOTILLO
Área de Medio Físico
Madrid, Spain

Sándor SZALAI
Hungarian Meteorological Service
Hungary

Tamás SZENTIMREY
Hungarian Meteorological Service
Hungary

John E. THORNES
School of Geography, Earth and Environmental Sciences
University of Birmingham
UK

Luiz Augusto TOLEDO MACHADO
Instituto Nacional de Pesquisas Espaciais – INPE
Centro de Previsão de Tempo e Estudos Climáticos – CPTEC
Brazil

Ole Einar TVEITO
Norwegian Meteorological Institute
Oslo, Norway

Zbigniew USTRNUL
Institute of Meteorology and Water Management
Kraków, Poland
University of Silesia, Sosnowiec
Poland

Hervé VAILLANT
LEGI
Joseph Fourier University
Grenoble, France

Riccardo VALENTINI
DISAFRI, Università degli Studi della Tuscia
Viterbo, Italy

F. VALERO
Dpto. Astrofísica y CC. de la Atmósfera, Facultad de CC Físicas
Universidad Complutense de Madrid
Madrid, Spain

Marco VIGLIOTTI
Dipartimento di Scienze Ambientali
Università degli Studi di Napoli
Caserta, Italy

Martin WEGEHENKEL
Institute of Landscape Systems Analysis, ZALF
Müncheberg, Germany

Steve WHITE
School of Geography, Earth and Environmental Sciences
University of Birmingham
UK

Index

S

T

U

W